THE TOKEN OF THE BOW

Enoch, Noah and Reclaiming the Rainbow

R. Lane Wright

The Nauvoo Press
Nauvoo, IL

© 2018 R. Lane Wright

All rights reserved. No part of this book may be reproduced in any form or by any means without permission in writing from the publisher, The Nauvoo Press, P.O. Box 134, Nauvoo, Illinois 62354.

Visit the publisher at www.TheNauvooPress.com

ISBN: **978-0-9863095-3-3**

Printed in the United States of America

10 9 8 7 6 5 4 3 2 1

THE TOKEN OF THE BOW

Enoch, Noah and Reclaiming the Rainbow

Table of Contents

THE BOW IN THE CLOUD...OUR STORY i

NOAH 1

BUILDING A BOAT 23

THE OPEN DOOR 51

THE FOUNTAINS OF THE GREAT DEEP 59

THE RAVEN AND THE DOVE 67

THE FOSSIL RECORD 71

HERE BE DRAGONS 101

RAIN OF HEAVEN 119

A FLOOD, A CITY, AND A GARDEN 131

THE AMAZING REACH OF COVENANTS 143

WHO SAYS THERE WAS A FLOOD ANYWAY? 151

SYMBOLS, SHADOWS AND STONE 167

LET THERE BE LIGHT 179

WHY ARE THERE SO MANY SONGS ABOUT RAINBOWS? 189

IS IT REALLY A RAINBOW AFTER ALL? 201

THIS SHALL BE A TOKEN... 207

THE COVENANT 211

AFTER THE FLOOD 219

MEAT FOR YOU.. 235

BAPTISM OF THE EARTH 243

WHO AM I?... 249

CONCLUSION 263

INDEX 267

To my family…all of you

The absence of family brought destruction to Noah's world.

The presence of family brought life to mine.

THE BOW IN THE CLOUD...OUR STORY

Since the time we were little children, we have heard the story of Noah and the Ark. We loved hearing of the great flood, and in particular about all the animals coming to Noah in pairs. Perhaps our fascination is that part of little children that loves animals, and water. We also love rainbows. This story has all the makings of a Grimm Fairy Tale. The seemingly innocuous story, with its beautiful images, and pretty pictures, yet somehow belying the dread and sorrow hiding just beneath the surface. While this may be perfectly fine for a children's fable, it does great injustice to a story as compelling and insightful as the story of Noah, and the great deluge, and the promise.

The most significant difference, is hopefully the most obvious: this story is true. As we will see in chapters to come, the epic tale of Noah and his splendid ark did in fact happen. More importantly, the message it conveys to us is every bit as important today as it was to a persecuted prophet thousands of years ago.

In Genesis, the Lord says; "This is the token of the covenant which I make between me and you, and every living creature that is with you, for perpetual generations: I do set my bow in the cloud, and it shall be for a token of a covenant between me and the earth."[1]

Isn't it interesting that the Lord is specific in his granting this covenant to "perpetual generations"? This sign was not just given for Noah, or those with him, or even those who knew him, but for everyone, even in the furthest reaches of time. It is a sign for us. It is a story for us.

But, how much of your time is consumed with worrying about a flood? I happen to live along the Mississippi River, and there are times when I do worry about a flood. It might inconvenience me, it might cause my basement to flood, it might even prevent me from getting where I want to go if roads are washed out. But, I do not worry that it will cover my house, or that it will cause me or my loved ones to be in danger, and I

---

[1] Genesis 9:12-13

certainly do not worry that it will cover the earth. So, this promise seems a little foreign to me. Does it to you?

There are other floods I do worry about. Psalms 18:4 says "The sorrows of death compassed me, and the floods of ungodly men made me afraid." Psalms 69:2 reads: "I sink in deep mire, where there is no standing' I am come into deep waters, where the floods overflow me." I can relate to these floods, they are very real to me. They trouble me, and those I love.

The Savior himself made reference to these spiritual floods in Luke 6:48 when he spoke about the man building his house upon the rock: "He is like a man which built an house, and digged deep, and laid the foundation on a rock: and when the flood arose, the stream beat vehemently upon that house, and could not shake it: for it was founded upon a rock."

The promise of the bow then, is also a reminder to us that the Lord has promised us, that by keeping his commandments, by being obedient, by building our house upon the rock, he will remember his promise and never allow us to perish.

The story of the bow is important because it is our story. The bow is a promise made to us and our families. We must understand it to take advantage of this amazing covenant.

And once we understand it, we must stand for it. In a world intent on drowning out a still, small voice with a cacophony of misinformation and misrepresentation, we are warned to remember, to stand, and to defend.

While others may be encouraged by evil forces to adopt the bow as a symbol of their own choices, we can be the guardians. As you will see, it is not simply a beautiful, colorful symbol, it is a token. A token of something important and valuable.

R. Lane Wright
2018

THE TOKEN OF THE BOW

Enoch, Noah and Reclaiming the Rainbow

NOAH

"Adam was made to open the way of the world, and for dressing the garden. Noah was born to save seed of everything, when the earth was washed of its wickedness by the flood; and the Son of God came into the world to redeem it from the fall...search the revelations of God; study the prophecies, and rejoice that God grants unto the world Seers and Prophets. They are they who saw the mysteries of godliness; they saw the flood before it came...they saw the stone cut out of the mountain, which filled the whole earth; they saw the Son of God come from the regions of bliss and dwell with men on earth; they saw the deliverer come out of Zion, and turn away ungodliness from Jacob; they saw the glory of the Lord when he showed the transfiguration of the earth on the mount; they saw every mountain laid low and every valley exalted when the Lord was taking vengeance upon the wicked; they saw truth spring out of the earth, and righteousness look down from heaven in the last days, before the Lord came the second time to gather His elect; they saw the end of wickedness on earth, and the Sabbath of creation crowned with peace; they saw the end of the glorious thousand years, when Satan was loosed for a little season; they saw the day of judgment when all men received according to their works, and they saw the heaven and the earth flee away to make room for the city of God, when the righteous receive an inheritance in eternity. And, fellow sojourners upon earth, it is your privilege to purify yourselves and come up to the same glory, and see for yourselves, and know for yourselves. Ask, and it shall be given you; seek, and ye shall find; knock, and it shall be opened unto you." [2]

When telling the story of Noah, we actually begin the story much earlier, with his great-grandfather Enoch. Another chapter in this book reviews the life of Enoch. So, we will begin with Enoch's son, Methuselah. We know Methuselah best from the modern

---

[2] *Evening and Morning Star*, August 1832

phrase "as old as Methuselah," a reference to the fact that he is the longest living individual we have reference to in scripture. Methuselah was a righteous man, and at one hundred years old, was ordained to the priesthood directly from the patriarch of all men, Adam.[3] He was a member of the group of righteous men who received their patriarchal blessings from Adam prior to his death. This group included Seth, Enos, Mahalaleel, Jared, Enoch and his son Methuselah.

As we will see, when the City of Enoch was taken, Methuselah was left behind, not due to the unrighteousness of Methuselah, but so "the covenants of the Lord might be fulfilled, which he made to Enoch; for he truly covenanted with Enoch that Noah should be of the fruit of his loins."[4]

A troubling phrase follows:

> "And it came to pass that Methuselah prophesied that from his loins should spring all the kingdoms of the earth (through Noah), and *he took glory unto himself.*"[5]

The way this is expressed indicates a certain amount of displeasure by the Lord. This feeling is only verified by the fact that, what

---

[3] Doctrine & Covenants 107:50
[4] Moses 8:2
[5] Moses 8:3, emphasis added

follows is the first famine recorded in scripture, and the fact that the Lord "cursed the earth with a sore curse…" Jewish legend says that it was the birth of Noah that caused the curse to come to an end.

The name Enoch gave to his son means "when he dies, judgment." We know much of Enoch's vision of the wickedness of men, and the coming of the Savior; and while much of it comes to us through restored scripture, fragments remain in the Old Testament. In particular, the book of Jude records Enoch's prophecy:

> "And Enoch also, the seventh from Adam, prophesied of these, saying, Behold, the Lord cometh with ten thousands of his saints, To execute judgment upon all, and to convince all that are ungodly among them of all their ungodly deeds which they have ungodly committed, and of all their hard speeches which ungodly sinners have spoken against him."[6]

Since Enoch knew of the coming of the flood, and the role his grandson would play in it, it is likely that Methuselah knew as well, but this was reaffirmed to him through the spirit of prophecy. From the apocrypha we learn that Methuselah was given his prophesy through a dream. When Methuselah awoke from this

---

[6] Jude 1:14-15

dream he was distressed. Perhaps it was this great distress that prompted Methuselah to ordain his grandson Noah when the boy was only ten years old. John Taylor said of the flood, and the resultant destruction of the wicked, that "Methuselah was so anxious to have it done that he ordained Noah to the Priesthood when he was ten years of age..." Methuselah died in the year of the great flood, he was 969 years old. Confirming the name given him by his father Enoch, for when he died, there was judgement.

Methuselah had a son, Lamech, who was the father of Noah. It is unclear what the meaning or origin of the name Lamech is, but a *"junior priest"* or possibly *"mighty youth* or *wild man"* are candidates.[7]

Joseph Smith said this of Lamech:

> The next great, grand Patriarch [after Enoch] who held the keys of the Priesthood was Lamech. "And Lamech lived one hundred and eighty-two years and begat a son, and he called his name Noah, saying, this same shall comfort us concerning our work and the toil of our hands because of the ground which the Lord has cursed."[8] The Priesthood continued from Lamech to Noah: "And God said unto

---

[7] M. Eliade, *The Forge and the Crucible*, 1962
[8] Genesis 5:28-29

> Noah, The end of all flesh is before me, for the earth is filled with violence through them and behold I will destroy them with the earth." [9],[10]

While we cannot say for a certainty who Noah's mother was, in Jasher 4:11 (an apocryphal work) it states that Lamech married Elisha's daughter Ashmua, granddaughter of Enoch.

Genesis 5:28-29 says, very matter-of-factly: "And Lamech lived an hundred eighty and two years, and begat a son: And he called his name Noah, saying, This same shall comfort us concerning our work and toil of our hands, because of the ground which the LORD hath cursed."

There is a tradition among the Jews that prior to the birth of Noah, when they planted wheat, oats would spring up, and that the oceans did not know their bounds, and would routinely flood their lands. They say that after the birth of Noah, all this stopped. The land behaved itself, and the crops were predictable. According to Jewish tradition, Noah invented the plow, the scythe, the hoe, and a number of other implements used for cultivating the ground. It is unclear how much of this is true, since we have no scriptural or prophetic basis for it, nonetheless it creates a beautiful picture of

---

[9] *Teachings of the Prophet Joseph Smith*, 171
[10] Genesis 6:13

how Noah has been venerated through the ages. It also falls neatly into place with the scripture in Genesis 6:8, "But Noah found grace in the eyes of the Lord."

While plants misbehaving may seem a little unbelievable, it is a great indicator into how people saw life in retrospect. The fallen world which came as after Eden was a very foreign place. Compared to the peaceful garden Adam had previously experienced, I'm sure his stories to his children would have made the world they lived in then seemed a very unpredictable and scary place. (See the chapter *After the Flood* for more insight)

There are things we do know about Noah. The Prophet Joseph Smith said that "Noah, who is Gabriel, … stands next in authority to Adam in the Priesthood; he was called of God to this office, and was the father of all living in his day, and to him was given the dominion. These men held keys first on earth, and then in heaven."

From Joseph, we also know that "as Noah was a preacher of righteousness he must have been baptized and ordained to the priesthood by the laying on of the hands."[11]

At 450 years old, Noah had a son, Japeth, and then 42 years later Shem, and eight years later Ham. Moses 8:13 utilized an

---

[11] *Teachings of the Prophet Joseph Smith*, Joseph Fielding Smith, 1976, 264

important phrase in describing Noah and his three sons. We first hear that they "hearkened unto the Lord," and they importantly "gave heed." As a result, they were "called the sons of God," (b'nei ha-Elohim).

This phrase, "sons of God," is often a distraction. It has been used by people through the years to imply that they were "fallen angels" or any number of other strange ideas. It is the writings of Enoch, repeated in the Book of Moses, which give the most convincing explanation: the beings who fell were not angels, but men who had become sons of God.

From the beginning, it tells us, mortal men could qualify as "sons of God," beginning with Adam.[12] How? By believing and entering the covenant.[13] Thus when "Noah and his sons hearkened unto the Lord, and gave heed … they were called the sons of God."[14] In short, the sons of God are those who accept and live by the law of God. When "the sons of men" (as Enoch calls them) broke their covenant, they still insisted on that exalted title: "Behold, we are the sons of God; have we not taken unto ourselves the daughters of men?"[15],[16]

---

[12] Moses 6:68
[13] Moses 7:1
[14] Moses 8:13
[15] Moses 8:21

One of Noah's sons, Shem is even thought by some to be the King of Salem himself, Melchizedek. They were certainly contemporaries (after the flood), and both are referred to with the descriptive phrase, "great high priest." "Melchizedek was such a great high priest"[17] and "Shem, the great high priest"[18]. In fact, Doctrine and Covenants 138 goes on to list Shem in the company of Abel, Seth, Noah, Abraham, and Isaiah (the name Melchizedek is conspicuously missing). Melchizedek may actually be considered a title. In Hebrew it translates to malki (מַלְכִּי), "My King" and sedek (צֶדֶק), "righteousness". Together, "My Divine King is Righteousness." While there certainly could have been two high priests, the honorific "great high priest" is enticing, as is the interpretation of Melchizedek as a title. The Times and Seasons[19] recognizes this connection stating, "Shem, who was Melchizedek".

Jewish tradition supports this contention, stating "Shem, or, as he is sometimes called, Melchizedek, the king of righteousness, priest of the Most High God…"[20] The Book of Jasher, Chapter 16, verse 11 and 12, says that "Adonizedek king of Jerusalem, the same was Shem, went out with his men to meet Abram and his people, with

---

[16] Hugh Nibley, *A Strange Thing in the Land: The Return of the Book of Enoch*, Part 8, Ensign, Dec 1976, 73
[17] Doctrine and Covenants 107:2
[18] Doctrine and Covenants 138:41
[19] *Times and Seasons*, Vol 5, No. 23, Dec 15, 1844, 746
[20] Ginsberg, *Legends of the Jews*, 233

bread and wine, and they remained together in the valley of Melech. And Adonizedek blessed Abram, and Abram gave him a tenth from all that he had brought from the spoil of his enemies, for Adonizedek was a priest before God."

In further support, much is known about Shem's youth, but little about his older years. With Melchizedek, the reverse is true, with nothing known of his youth, but much about his later years, and service as the great High Priest. It is encouraging to think of Shem, joining his ancestor Enoch, taking the righteous citizens of his city with him.

Noah's sons had families. Among these families were daughters, granddaughters to Noah. Scripture says that "those daughters were fair." Those who sought to marry them are identified by a unique distinction. Where Japeth, Shem and Ham are called "sons of God" (living under the covenant), those who marry their daughters are called the "sons of men" (living outside the covenant). Their sin was compounded in the statement at the close of verse 14, saying they "took them wives, *even as they chose*." They had stopped listening to God.

Compare the rendition of this event in the Bible, with what was restored through the Prophet Joseph in the Book of Moses:

The Book of Genesis: "And it came to pass, when men began to multiply on the face of the earth, and daughters were born unto them, That the *sons of God* saw the daughters of men that they were fair; and they took them wives of all which they chose."[21]

The Book of Moses: "And Noah and his sons hearkened unto the Lord, and gave heed, and they were called the sons of God. And when these men began to multiply on the face of the earth, and daughters were born unto them, the *sons of men* saw that those daughters were fair, and they took them wives, even as they chose."[22]

Notice that the emphasis has shifted from the "Sons of God" as the offender, to the "sons of men." And, through this distinction, a clearer understanding of who these groups is possible.

"Because the daughters of Noah married the sons of men contrary to the teachings of the Lord, his anger was kindled, and this offense was one cause that brought to pass the universal flood… The daughters who had been born, evidently under the covenant, and were the daughters of the sons of God, that is to say of those

---

[21] Genesis 6:1-2 (emphasis added)
[22] Moses 8:13-14 (emphasis added)

who held the priesthood, were transgressing the commandment of the Lord and were marrying out of the Church. Thus, they were cutting themselves off from the blessings of the priesthood contrary to the teachings of Noah and the will of God…"[23]

Joseph Fielding Smith said this of the daughters of Noah:

> "Because the daughters of Noah married the sons of men contrary to the teachings of the Lord, his anger was kindled, and this offense was one cause that brought to pass the universal flood…The daughters who had been born, evidently under the covenant, and were the daughters of the sons of God, that is to say of those who led the priesthood, were transgressing the commandment of the Lord and were marrying out of the Church. Thus they were cutting themselves off from the blessings of the priesthood contrary to the teaching of Noah and the will of God…"

He continues,

> "Today there are foolish daughters of those who hold this same priesthood who are violating this commandment and marrying the sons of men; there are also some of the sons of those who hold the priesthood who are marrying the

---

[23] Pearl of Great Price Student Manual – Religion 327

daughters of men. All of this is contrary to the will of God just as much as it was in the days of Noah."[24]

It was this sequence of events—living outside the covenant, and refusing to listen to the Lord—that led to the Lord saying to Noah that the time had come. He instructed Noah to warn the people, that the flood, promised so long ago would come. The immediate impact of the Lord's displeasure however was the limit he placed on their time. The consequence of the fact that the Spirit would not be with them is two-fold: first that their lives would be limited to 120 years, and if they did not repent, the floods would come.

If limiting their lifespan does not cause them to be more obedient, then more drastic measures will have to be taken. The effect is not immediate, but it is very real.

---

[24] Joseph Fielding Smith, *Answers to Gospel Questions*, 1:136-137

How Long Did They Live?	
Adam	930
Seth	912
Enosh	905
Cainan	910
Mahalalel	895
Jared	962
Enoch	365
Methuselah	969
Lamech	777
Noah	950
Genesis 6:3 (120 years limit)	
Shem	600
Arphaxad	438
Salah	433
Eber	464
Peleg	239
Reu	239
Serug	230
Nahor	148
Terah	205
Abraham	175
Ishmael	137
Isaac	180
Jacob (Israel)	147
Joseph	110
Moses	120
Joshua	110
Lehi	75
Nephi	75

What follows is a reference that has caused many to wonder. The scriptures say that "in those days there were giants on the earth, and they sought Noah to take away his life."[25]

The word translated as "giants" in this verse is the Hebrew word nephilim. There has been much debate over the meaning of this word. Some believe it comes from the Hebrew verb naphal, while others claim that it is from the Aramaic noun naphil. These individuals are further described in Hebrew as gibborim ("mighty men").

The nephilim were mentioned again when the spies returned from their exploratory mission of the land of Canaan. These spies reported that Ahiman, Sheshai, and Talmai (descendants of Anak, progenitor of the Anakim) dwelt in Hebron. They also stated, "...the people be strong that dwell in the land, and the cities are walled, and very great: and moreover we saw the children of Anak there."[26]

The chapter concludes with ten of the spies giving "a bad report" trying to convince the Israelites that they could not conquer the land:

> "And they brought up an evil report of the land which they had searched unto the children of Israel, saying, The land,

---

[25] Moses 8:18
[26] Numbers 13:28

> through which we have gone to search it, is a land that eateth up the inhabitants thereof; and all the people that we saw in it are men of a great stature. And there we saw the *giants*, the sons of Anak, which come of the *giants*: and we were in our own sight as grasshoppers, and so we were in their sight."[27]

The Hebrew נפלים, (the nephilim of Moses 8:18, Genesis 6:4 and Numbers 13:33) is translated as "giants." We get this rendition largely due to an original translation into Greek, the Septuagint, of the Hebrew Torah. There nephilim is translated as γιγαντες (gigantes), hence "giant." However, gigantes is literally "earth-born". The Hebrew term nephilim is more accurately translated as "he fell," as in "apostatized or fallen from the true religion."

And these "mighty men—men of renown" (גברים), the gibborim, translates best as "conquerors or heroes". In the Greek translation, it also represents anthropoi onomastoi (ανθρωποι ονομαστοι), or a unique phrase, "twice named"; named once with the name they received from their fathers, and again with the name they received for their daring exploits. This is not to say that there were never "giants", taller than average men, on the earth. But, we don't need to reach for fallen angels, or alien giants to understand the Nephilim of Noah.

---

[27] Numbers 13:32-33 (*Emphasis Added*)

In spite of all of the evil in the world, God's love for his children had still not caused him to leave them hopeless. Following the pattern he has always followed, the Lord protected Noah, "ordained Noah after his own order, and commanded him that he should go forth and declare his Gospel unto the children of men, even as it was given unto Enoch."[28] Noah was given the authority, and the calling to preach to the people, to call them to repentance, just as his great-grandfather had done. Enoch was provided a vision, and a pattern to teach the people. Noah was given this same instruction as the process the people must embrace in order to have a terrible curse lifted from them. The process was the same then as now, regardless of their level of disobedience. They were called to repent. And what follows is all to typical.

They were not content to simply disobey, or refuse to repent, they tried to justify to Noah (and therefore God), why their behavior was acceptable. They explained to Noah how they were actually doing just fine. They were marrying, having children, and these children were doing great things.

They boasted of "eating and drinking, and marrying and giving in marriage."[29] No mention of God, or of the Covenant. This comment by the people who refused to listen to Noah, becomes

---

[28] Moses 8:19
[29] Matthew 24:38

an example used by Savior when He is describing the time when He will return. He said to His disciples (in reply to the question: "Tell us, when shall these things be? and what shall be the sign of thy coming, and of the end of the world?"[30]) : "For as in the days that were before the flood they were eating and drinking, marrying and giving in marriage, until the day that Noe [Noah] entered into the ark, And knew not until the flood came, and took them all away..."[31] They were so preoccupied with their own lives, with the everyday events (eating and drinking), and with their own version of the covenant, a simple business arrangement, (marrying and giving in marriage) that they had no time for the Lord, or use for Him in their lives.

The Jewish *Book of Jasher* shares an interesting perspective on these events:

> "And after the lapse of many years, in the four hundred and eightieth year of the life of Noah, when all those men, who followed the Lord had died away from amongst the sons of men, *and only Methuselah was then left,* God said unto Noah and Methuselah, saying, Speak ye, and proclaim to the sons of men, saying, Thus saith the Lord, return from your evil ways and forsake your works, and the Lord will repent

---

[30] Matthew 24:3
[31] Matthew 24:38-39 and Luke 17:27

of the evil that he declared to do to you, so that it shall not come to pass. For thus saith the Lord, Behold I give you a period of one hundred and twenty years; if you will turn to me and forsake your evil ways, then will I also turn away from the evil which I told you, and it shall not exist, saith the Lord. *And Noah and Methuselah spoke all the words of the Lord to the sons of men*, day after day, constantly speaking to them. But the sons of men would not hearken to them, nor incline their ears to their words, and they were stiffnecked."[32]

It is interesting that in this rendition, Methuselah is beside Noah, the last in the line of righteous men, proclaiming the gospel, and calling the people to repentance. The phrase "and only Methuselah was then left…" is heart rending.

With one last call to a sinful world to repent "and be baptized in the name of Jesus Christ"[33], Noah reveals to the people, again, that the flood will come upon them. They may very well have heard of the prophesies of Enoch regarding a flood, since he had seen it and written it in his book, but now it was real, now it would affect them. Some of the most painful words of scripture are then recorded, "nevertheless they hearkened not." An all-to-familiar refrain, heard even in our day, "they hearkened not,"

---

[32] Jasher 5:6-10
[33] Moses 8:24

the people refused to head the warnings of a loving God, voiced by his chosen messenger, the Prophet.

Those who do not have access to restored scripture now must do linguistic gymnastics as they attempt to reconcile the words "And it repented the Lord that he had made man on the earth, and it grieved him at his heart." If the Lord had a need to repent, did that mean he had made a mistake? The Hebrew word used here, that is translated "repent" is nacham (נָחַם). This word implies a change in direction, or consoling or comforting. Perhaps the Lord decided it was time to change directions, or was simply saying he wished to console Noah, or humanity.

Through the Prophet Joseph, what was once a subject of discussion, and confusion, was restored to its consistent, compassionate and understandable form. Moses 8:25 was restored to read, "And it repented Noah, and his heart was pained that the Lord had made man on the earth, and it grieved him at the heart." Here we see that it was Noah who was sorry, who needed consolation and comforting. Why? As we read previously, his granddaughters had "sold themselves." They had taken "sons of men" as husbands, turning their back on the covenants of the Lord.

What can cause a grandfather to wish all of humanity had never been created? The loss of his grandchildren. The judgement of the world came because the world had not only turned its back on God, it had now rejected the one vehicle the Lord had provided to bring them back to him, even when they strayed...family.

As a result of the unwillingness of the people to repent, and the loss of families, the Lord said "I will destroy man whom I have created, from the face of the earth, both man and beast and the creeping things, and the fowls of the air; for it repenteth Noah that I have created them..."

There is one final straw for the Lord. As if it were not enough that they have rejected the Lord, His covenant, their own families, His commandments and the Savior that would be provided for them, they also rejected the one man the Lord trusted to teach and bring them back to Him, "for they have sought his life."[34]

One last time God pronounces his judgement, "The end of all flesh is come before me...I will destroy all flesh from off the earth."[35]

Now it's time to build a boat.

---

[34] Moses 8:26
[35] Moses 8:30

Berkeley Wright, 2017

BUILDING A BOAT

President Spencer W. Kimball explained that when Noah built the ark, "there was no evidence of rain and flood. . . . His warnings were considered irrational. . . . How foolish to build an ark on dry ground with the sun shining and life moving forward as usual! But time ran out. The ark was finished. The floods came. The disobedient and rebellious were drowned. The miracle of the ark followed the faith manifested in its building"[36]

We cannot really call Noah's grand vessel a boat, since a boat requires a means of propulsion (either sail, oars or engine), but for the sake of this volume, we will stretch the definition a bit, and call it a boat. However, in scripture, the vessel Noah was commanded to build was called an ark. In Hebrew, ark is tebah (תֵּבָה). This is

---

[36] Faith Precedes the Miracle, 1972, 5–6

the first ark in scripture, but hardly the last. This ark was followed by the ark built by Moses' mother[37]. The ark built to carry the artifacts provided by the Lord, as the children of Israel wandered in the wilderness, is confusing. While translated as ark, it is actually translated from a different Hebrew word, aron (אָרוֹן). This is the term used to describe the ark of the covenant. This term also appears in Genesis[38], but here describes a coffin. Each time the ark of the covenant is mentioned, in Exodus, Numbers, Joshua, Samuel and Chronicles, it is aron that is being referenced. When I use the term ark hereafter, I am referring to tebah, not aron.

---

[37] Exodus 2:3
[38] Genesis 50:26

In each case, an ark (tebah) was built as a means of protection and salvation. Noah's ark saved a righteous family from God's judgment on humanity. Moses' ark saved the man who would lead a nation to safety after God's judgment on Egypt.

Joseph Smith said of the Ark: "The construction of the first vessel was given to Noah, by revelation. The design of the ark was given by God, 'a pattern of heavenly things'."[39]

The Lord provided instruction to Noah in building his great ark:

> "Make thee an ark of gopher wood; rooms shalt thou make in the ark, and shalt pitch it within and without with pitch. And this is the fashion which thou shalt make it of: The length of the ark shall be three hundred cubits, the breadth of it fifty cubits, and the height of it thirty cubits. A window shalt thou make to the ark, and in a cubit shalt thou finish it above; and the door of the ark shalt thou set in the side thereof; with lower, second, and third stories shalt thou make it."[40]

As I discussed in the chapter on Noah, historically Noah is seen as the inventor of tools, and even of farming (being the creator of the plow). But, just as Nephi needed instruction on building his boat,

---

[39] Joseph Fielding Smith, *Teachings of the Prophet Joseph Smith*, 1976, 251
[40] Genesis 6:14-16

Noah would need help. We begin with the material to build his vessel.

The Hebrew word gopher is used only once in the Bible…when God commanded Noah to "make yourself an ark of gopher wood"[41]. Because no one today knows what "gopher wood" is, although Noah obviously knew, the King James Version simply transliterates the Hebrew and leaves it as "gopher wood." When you say 'gopher wood', you are actually combining a Hebrew word and an English word to make up the phrase. The Septuagint (the Greek Old Testament) renders the phrase as "squared beams," and the Latin Vulgate simply says "planed wood."

It is possible that gopher wood is a species of tree that did not survive the flood. It may be one of the many fossil trees we find that no longer exist. Noah would likely have been in the vicinity of Adam Ondi Ahman, which also coincides to the near center of the land mass that was initially contiguous, and called Rodinia (discussed further in *The Fountains of the Great Deep* chapter).

Rodinia predates the supercontinent Pangea, and is the name given to the first single land-mass, prior to the separation and collisions that created the great seas, and mountains.

---

[41] Genesis 6:14

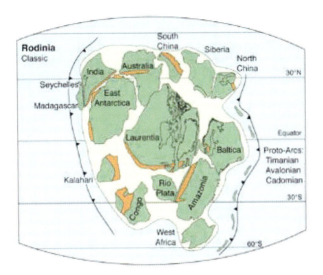

Figure 1: Rodinia

Current day Missouri would have been located near the center of this land mass, in the area labeled, Laurentia. This map was created using core samples and paleomagnetic data which identify the jigsaw puzzle pieces which would have existed prior to the sediments which enlarged the continents during and after the flood. This would have then been the location of Adam ondi Ahman, where Adam blessed his posterity, and assuming that there was little need to wander too far from Eden after Adam's expulsion, was likely where the original garden home would have been located. This is on or near the equator, and so would have been a very tropical location. There were no large mountain ranges yet, since there had been no continental collisions to create them.

There were huge forests of trees in the area of Missouri and Illinois that were buried by the same flood waters that lifted the ark. The largest underground fossil forest ever found was discovered near modern-day Galatia, Illinois. This forest extends more than 100 miles. The trees that make up this forest were 100 feet tall and more, and six feet wide at the base, retaining that width to their height. It is worth noting that paleobiologists studying the area believe that these trees were killed by floods. These trees no longer grow in Illinois, or anywhere else.[42]

Figure 2: The dark mass is a coal seam; the lighter shale above is interrupted by a fossil tree stump.

---

[42] W. Barksdale Maynard, "An Underground Fossil Forest Offers Clues on Climate Change," *The New York Times,* April 30, 2012

Some researchers believe the word gopher doesn't refer to a species of wood at all; rather, it refers to a process utilized to prepare the wood in the ark's construction. This is seen in the Septuagint's translation, "squared beams." Some archaeologists have suggested that gopher may have referred to a lamination process, made necessary by the enormous size of the ark.

To add more speculation to the meaning of "gopher wood," there is even disagreement as to the true spelling of the Hebrew word. Due to the similarity between a 'g' (gimel, ג) and a 'k' (kaf, כ) in the Hebrew alphabet (both letters resemble a backwards 'c'), some scholars have proposed that the first letter in the word gopher was inadvertently switched by a scribe, and that the word should actually be kopher (כֹּפֶר), a Hebrew word meaning "pitch." If this scribal-error theory is correct, Genesis 6:14 would read, "Make yourself an ark of pitched (waterproofed) wood; make rooms in the ark, and cover it inside and outside with pitch."

The Syrian Christians have a long-standing tradition (which has spread to Egypt, Greece, Romania, Bulgaria and other eastern groups) that God ordered Noah to make the first semantron and mallet and to strike the mallet against the semantron three times every day while the ark was being built. According to the story, God told him: "Make for yourself a bell of box-wood, which is not liable to corruption, three cubits long and one and a half wide, and also a mallet from the same wood. Strike this instrument three separate times every day: once in the morning to summon the hands to the ark, once at midday to call them to dinner, and once in the evening to invite them to rest".

WOODEN SEMANTRON.

The semantra is struck when it is time to summon the people to public prayer. Tradition also links the sound of the wood to the wood of the Garden of Eden that caused Adam to fall when he plucked its fruit, and to the nailing of Christ to the wood of the cross, to atone for Adam's transgression[43].

As the Lord continued to provide design guidance to Noah, he gives him the dimensions of the ark. The measurements he is given are 300 cubits long, by 50 cubits wide and 30 cubits high. A cubit in the Old Testament was generally about 17.5 inches. However, an Egyptian royal cubit measured about 20.5 inches. Since Moses was educated in Egypt we must allow for the possibility that the longer measurement was meant here. The Ark, therefore, could

---

[43] John O'Brien, *A History of the Mass and Its Ceremonies in the Eastern and Western Church*, 1879, 148-49

have measured from 437 feet to 512 feet in length! It was not until the late 19th century that a ship anywhere near this size was built. Beyond the amazing size, there are inherent characteristics of these measurements which we should consider. The Ark had a ratio (length x width x height) of 30 x 5 x 3. According to ship-builders, this ratio represents an advanced knowledge of ship-building since it is the optimum design for stability in rough seas. The Ark, as designed by God, was virtually impossible to capsize. It would have to have been tilted over 90 degrees in order to capsize.

Figure 3: Currier & Ives Noah's Ark

Very little in the way of detail is provided in the shape of the ark. But, the use of the Hebrew term 'tebah" may provide some insight. The word tebah is only used in the Bible to describe Noah's ark, and Moses' basket or ark. It is not likely that this was a typical ship or boat, since there are descriptive terms that would more accurately describe that kind of vessel. Ship or boat is the Hebrew word oniyyah (אֳנִיָּה). Tebah describes a box, chest or basket. Therefore, rather than a vessel with a keel, as a ship would have, we can expect a vessel with a square bottom, similar to a barge, that would ride low in the water. This has the advantages of being stronger on dry land (without the need for a complicated dock), as well as being stable at sea, which would be helpful since it would have no method of propulsion, and no rudder to steer.

The Lord instructed that it would be covered in pitch, both inside and out[44]. Since the kind of pitch used today is derived from oil, which is a fossil product, the pitch used by Noah would have been of a different method or type. It is interesting that the Hebrew word "kopher," is usually translated in scripture as "atonement," and is here translated as pitch, or "a covering." Pitch is usually understood to be a black glue-like substance left behind when coal tar is heated or distilled. It belongs to the same family of substances as asphalt or bitumen. Today, it is largely produced by heating coal. But coal tar and petroleum are not the only source for pitch.

---

[44] Genesis 6:14

For at least one thousand years, prior to the introduction of petroleum tar and pitch, pitch-making in Europe assisted in the construction of great wooden sailing ships which figured so prominently in history.[45]

The first step in this process was to obtain resin from the pine trees, which at that time grew in dense forests throughout Europe. A herringbone pattern of cuts was gouged into the tree trunk and as the resin ran down the grooves, it was collected in a pot at the base of the tree, and heated.

When the resin had finished flowing, the trees were chopped down, covered in soil or ash, and burned slowly to produce a lightweight black pure form of carbon called charcoal. The last step in the process of making pitch was to add the powdered charcoal to the boiling pine resins. Different proportions of charcoal would produce pitch of different properties. It was this pitch which was used to waterproof the large ocean-going wooden ships, and may have been the 'kopher' Noah referred to.

The Flood and the Ark are great symbols of salvation for those who put their trust in God. In order to fully grasp this idea, we

---

[45] Gamble, Thomas. Ed. "How The Famous "Stockholm Tar" of Centuries of Renown Is Made," a 1914 Report. Naval Stores: History, Production, Distribution and Consumption, *Savannah: Weekly Naval Stores Review*, 1921. 47

need to look deep into the Hebrew of it all. In the original text, we find a particular verb כפר "kopher" (to atone) which shares the same root of the word used for "Yom Kipur" which is the day of Atonement. This verb is used to express both the idea of "to cover with pitch" and "atone for sin". This simple practical command made by God, sounds in Hebrew almost like a theological statement. Through this understanding of the Hebrew, it becomes immediately clear that the story of Noah is essentially a story of redemption and atonement.

The matter of the window in the ark is a complicated one, made even more so by the inconsistencies in the Hebrew translations. In Genesis 6:16, it says:

> "A window shalt thou make to the ark, and in a cubit shalt thou finish it above; and the door of the ark shalt thou set in the side thereof; with lower, second, and third stories shalt thou make it."

The word translated "window" here is the Hebrew word tsohar (צֹהַר), and means "noon" or "midday." It occurs next in Genesis 43:16, where it is translated as "noon":

> When Joseph saw Benjamin with them, he said to the steward of his house, "Take these men to my home, and slaughter an

animal and make ready; for these men will dine with me at noon."

The King James Version translates tshohar word as "noon" 11 times, "noonday" 9 times, "day" 1 time, "midday" 1 time, "noontide" 1 time (along with a second word,) and "window" just this one time. The word tsohar means "noon," not "window."

The same window is mentioned again in Genesis, but this time it is a different Hebrew word. The reference is Genesis 8:6:

"So it came to pass, at the end of forty days, that Noah opened the window of the ark which he had made."

This time, the Hebrew word translated "window" is challown (חַלּוֹן), which is the common word for a "window," but literally means a "piercing" in the wall. It occurs 31 times, and is translated "window" all 31 times in the King James Version. This is the only time that it is used for a "window" on a boat, however.

Therefore, we are left with two words to figure out what exactly this "window" was that Noah made in the ark, and opened to let the birds out. It is on a ship, it is pierced through the ship, and it can be described as a "noon." This last fact would seem important to guide us. It seems logical that this opening was in the roof of the ark, not the wall, where we would normally think of a window as being located. This "noon" Noah was to make, this "piercing" in the ark, was actually a hatch located in the top of the ark. A hatch, being straight overhead, could easily be called a "noon." It is not a window, but it certainly is a "piercing" in the ark, one straight overhead rather than to the side, as a window would be. A hatch would be an obvious thing to open in order to let a raven or a dove out of the ark, and to receive it back in.

A window would make little sense in the side of a ship created to ride out the worst storm in history. From a window, Noah could have surveyed the earth's surface himself, not needing a bird to do it. But a hatch, about 18 inches around (one cubit), would be much safer than a window, and would be the kind of thing one would expect to find on a boat like the ark.

The footnote in the Latter-day Saint scriptures for Genesis 6:16 contains an intriguing detail that encourages additional consideration. It states: "HEB tsohar; some rabbis believed it was a precious stone that shone in the ark." And then provides a cross-reference to Ether 2:23 where the brother of Jared receives instruction that he cannot have a window, but instead creates sixteen small stones, which the Lord illuminates in a dramatic way.

The connection to a rabbinic tradition, referenced in the Talmud, is amazing. Rather than translating tsohar as either a window or a hatch, it refers to a "light". The implication is that Noah gathered precious stones and jewels to use as a lantern. The Talmud states that when the stones shone dimly, it was daytime outside the ark, and when they shone brightly it was night. This allowed Noah not only to see, but to know the proper time for feeding of the animals in the ark. It is thought that this stone, called Jacinth, was a red garnet.

> "Go thou unto Phison, and take from thence a precious stone, and fix it in the ark to illuminate you: with the measure of a cubit (or span) shalt thou complete it above." [46],[47]

The river Phison was the first of four rivers named in Genesis 2:10-14 that emanate from the river that watered the garden of Eden. "The name of the first is Phison: this is the one that compasses the entire land of Havilah, where there is gold".[48] This would further verify that Noah did not travel far from the garden for his home.

The midrash by Pirke de Rabbi Eliezer records that Rabbi Shemiah taught, "The Holy One, blessed be He, showed Noah with a finger and said to him, Like this and that shalt thou do the ark."[49] According to this rabbinic legend, Noah saw the finger of the Lord when he was instructed on how to build the ark. This is familiar

---

[46] Genesis 6:16
[47] *The Targum of Jonathan Ben Uzziel*
[48] Genesis 2:11
[49] Gerald Friedlander, Trans., Pirke de Rabbi Eliezer, New York: Bloch Publishing Co., 1916, 164

enough to the experience of the Brother of Jared to ring true to Latter-day Saints. It's even possible that this event served as an inspiration to the Brother of Jared when he was faced with a similar situation. He could simply have been doing what he knew Noah had done before him.

In researching garnet stones in the area where Noah likely built the ark, it was interesting to discover that the largest garnet ever discovered[50], was found in New York in 1885. At the time of its discovery it was called the "subway garnet" since local media reports indicated it was found while cleaning a subway tunnel. In fact, it was discovered while excavating a sewer line, but "sewer garnet" does not have quiet the appeal. This garnet was the size of a bowling ball, and was used for years as a door-stop in the public works office. It is now located in the American Museum of Natural History. "It's rare in the sense of the size and perfection of it," said George Harlow, the curator for the earth sciences department at the museum. "They're usually all broken up at that size."

---

[50] The Subway Garnet, *The New York Times*, December 24, 2015

BUILDING A BOAT 39

Figure 4: The nine-pound, 6" diameter, "Subway Garnet"

To give some perspective on the size of the ark, and its ability to house so many "kinds" of animals. Let's do a little math. As discussed earlier, the ark could have been up to 550 feet long, 91.7 feet wide and 55 feet high. But how much storage space does this amount to? Well, 550 x 91.7 x 55 = 2,773,925 cubic feet. (If we take the smallest measurement of cubit, 17 inches, we end up with 1,278,825 cubic feet). Of course, not all of it would have been free space. The ark had three levels[51] and a lot of rooms[52], the walls of which would have taken up space. Nevertheless, it has been

---

[51] Genesis 6:16
[52] Genesis 6:14

calculated that a little more than half (54.75%) of the 2,773,925 cubic feet could store 125,000 sheep-sized animals. With sheep being a relatively average sized animal, we begin to realize that a vessel of this size could carry the large menagerie described in scripture.

So, which animals were included?

> "And of every living thing of all flesh, two of every sort shalt thou bring into the ark, to keep them alive with thee; they shall be male and female. Of fowls after their kind, and of cattle after their kind, of every creeping thing of the earth after his kind, two of every sort shall come unto thee, to keep them alive."[53]

The word translated as 'kind' is the Hebrew word "min" (מִין). It's literal meaning is "a portion of", it is translated as kind, or even species. This word appears only 31 times in the Old Testament. Due to the amazing complexities of DNA, a limited number of individuals can result in a broad range of species through hybridization. To many this may sound like evolution at work, and perhaps in some sense it is. But the many different interpretations of this concept make it problematic to use as a

---

[53] Genesis 6:19-20

definition. (Review the chapter, *The Fossil Record* for a more detailed discussion of this topic.)

For our purposes here, I will simply state that through the principle of genetic "kinds" or even extending it to representative "species", and the amazing power of DNA, a large variety of organisms can result from a more limited number of passengers on the ark. The miracle is not that so many animals will fit in one ark, but that so much information was instilled in these organisms, giving them, by design, an amazing adaptability.

So, Noah was given everything he needed to fulfill the Lord's request. He had the materials, the tools, the design, and presence of the Lord. What more could he possibly need? His family. At the heart of this whole process has been the preservation of family. This would be no exception. Not only would the ark save Noah and his wife (the new Adam and Eve), but also his three sons; Japeth, Shem and Ham, and their wives. Unfortunately, they will not be accompanied by Noah's granddaughters. As we already learned, they had sold themselves to the world. But, as with everyone else, the door was opened to them to repent and be baptized.

Until the door was shut.

42　THE TOKEN OF THE BOW

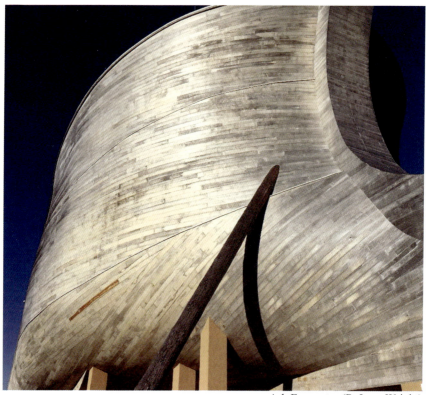

Ark Encounter (R. Lane Wright)

CHAPTER END NOTES

In addition to providing instructions to Noah, the Lord has also provided detailed instructions on other construction projects. Because I am a fan of the space program, I can't help but think of these projects as great leaps forward. Just as the space program provided benefits to subsequent generations, the instruction provided by the Lord on the ark, these temples, and other ships would have benefited generations that followed in amazing ways.

Instructions for the building of Solomon's Temple were given to David by the Lord:

> "Then David gave to Solomon his son the pattern of the porch, and of the houses thereof, and of the treasuries thereof, and of the upper chambers thereof, and of the inner parlours thereof, and of the place of the mercy seat, And the pattern of all that he had by the spirit, of the courts of the house of the LORD, and of all the chambers round about, of the treasuries of the house of God, and of the treasuries of the dedicated things: Also for the courses of the priests and the Levites, and for all the work of the service of the house of the LORD, and for all the vessels of service in the house of the LORD. He gave of gold by weight for things of gold, for all instruments of all manner of service; silver also for all instruments of silver by weight, for all instruments of every kind of service: Even the weight for

the candlesticks of gold, and for their lamps of gold, by weight for every candlestick, and for the lamps thereof: and for the candlesticks of silver by weight, both for the candlestick, and also for the lamps thereof, according to the use of every candlestick. And by weight he gave gold for the tables of shewbread, for every table; and likewise silver for the tables of silver: Also pure gold for the fleshhooks, and the bowls, and the cups: and for the golden basons he gave gold by weight for every bason; and likewise silver by weight for every bason of silver: And for the altar of incense refined gold by weight; and gold for the pattern of the chariot of the cherubims, that spread out their wings, and covered the ark of the covenant of the LORD.

"All this, said David, the LORD made me understand in writing by his hand upon me, even all the works of this pattern.

"And David said to Solomon his son, Be strong and of good courage, and do it: fear not, nor be dismayed: for the LORD God, even my God, will be with thee; he will not fail thee, nor forsake thee, until thou hast finished all the work for the service of the house of the LORD."[54]

---

[54] 1 Chronicles 28:11-20

Instructions for building the Tabernacle in the wilderness:

"Moreover thou shalt make the tabernacle with ten curtains of fine twined linen, and blue, and purple, and scarlet: with cherubims of cunning work shalt thou make them. The length of one curtain shall be eight and twenty cubits, and the breadth of one curtain four cubits: and every one of the curtains shall have one measure. The five curtains shall be coupled together one to another; and other five curtains shall be coupled one to another.

"And thou shalt make loops of blue upon the edge of the one curtain from the selvedge in the coupling; and likewise shalt thou make in the uttermost edge of another curtain, in the coupling of the second. Fifty loops shalt thou make in the one curtain, and fifty loops shalt thou make in the edge of the curtain that is in the coupling of the second; that the loops may take hold one of another. And thou shalt make fifty taches of gold, and couple the curtains together with the taches: and it shall be one tabernacle.

"And thou shalt make curtains of goats' hair to be a covering upon the tabernacle: eleven curtains shalt thou make. The length of one curtain shall be thirty cubits, and the breadth of one curtain four cubits: and the eleven curtains shall be all of one measure. And thou shalt couple five curtains by themselves, and six curtains by themselves, and shalt double the sixth curtain in the forefront of the tabernacle. And thou shalt make fifty loops on the edge of the one curtain that is outmost in the coupling, and fifty loops in the edge of the curtain which coupleth the second.

"And thou shalt make fifty taches of brass, and put the taches into the loops, and couple the tent together, that it may be one. And the remnant that remaineth of the curtains of the tent, the half curtain that remaineth, shall hang over the backside of the tabernacle. And a cubit on the one side, and a cubit on the other side of that which remaineth in the length of the curtains of the tent, it shall hang over the sides of the tabernacle on this side and on that side, to cover it.

"And thou shalt make a covering for the tent of rams' skins dyed red, and a covering above of badgers' skins. And thou shalt make boards for the tabernacle of shittim wood standing up. Ten cubits shall be the length of a board, and a cubit and a half shall be the breadth of one board. Two tenons shall there be in one board, set in order one against another: thus shalt thou make for all the boards of the tabernacle. And thou shalt make the boards for the tabernacle, twenty boards on the south side southward. And thou shalt make forty sockets of silver under the twenty boards; two sockets under one board for his two tenons, and two sockets under another board for his two tenons. And for the second side of the tabernacle on the north side there shall be twenty boards: And their forty sockets of silver; two sockets under one board, and two sockets under another board.

"And for the sides of the tabernacle westward thou shalt make six boards. And two boards shalt thou make for the corners of the tabernacle in the two sides. And they shall be coupled together beneath, and they shall be coupled together above the head of it unto one ring: thus shall it be for them both; they shall be for the two corners. And they shall be eight boards, and their sockets of silver, sixteen sockets; two sockets under one board, and two sockets under another board. And thou shalt make bars of shittim wood; five for the boards of the one side of the tabernacle, And five bars for the boards of the other side of the tabernacle, and five bars for the boards of the side of the tabernacle, for the two sides westward.

"And the middle bar in the midst of the boards shall reach from end to end. And thou shalt overlay the boards with gold, and make their rings of gold for places for the bars: and thou shalt overlay the bars with gold. And thou shalt rear up the tabernacle according to the fashion thereof which was shewed thee in the mount. And thou shalt make a vail of blue, and purple, and scarlet, and fine twined linen of cunning work: with cherubims shall it be made: And thou shalt hang it upon four pillars of shittim wood overlaid with gold: their hooks shall be of gold, upon the four sockets of silver. And thou shalt hang up the vail under the taches, that thou mayest bring in thither within the vail the ark of the testimony: and the vail shall divide unto you between the holy place and the most holy.

"And thou shalt put the mercy seat upon the ark of the testimony in the most holy place. And thou shalt set the table without the vail, and the candlestick over against the table on the side of the tabernacle toward the south: and thou shalt put the table on the north side. And thou shalt make an hanging for the door of the tent, of blue, and purple, and scarlet, and fine twined linen, wrought with needlework. And thou shalt make for the hanging five pillars of shittim wood, and overlay them with gold, and their hooks shall be of gold: and thou shalt cast five sockets of brass for them."[55]

The Brother of Jared's instructions to build barges:

"And the Lord said: Go to work and build, after the manner of barges which ye have hitherto built. And it came to pass that the brother of Jared did go to work, and also his brethren, and

---

[55] Exodus 26

built barges after the manner which they had built, according to the instructions of the Lord. And they were small, and they were light upon the water, even like unto the lightness of a fowl upon the water.

"And they were built after a manner that they were exceedingly tight, even that they would hold water like unto a dish; and the bottom thereof was tight like unto a dish; and the sides thereof were tight like unto a dish; and the ends thereof were peaked; and the top thereof was tight like unto a dish; and the length thereof was the length of a tree; and the door thereof, when it was shut, was tight like unto a dish.

"And it came to pass that the brother of Jared cried unto the Lord, saying: O Lord, I have performed the work which thou hast commanded me, and I have made the barges according as thou hast directed me."[56]

Nephi's receives instructions to build a ship:

"And it came to pass that the Lord spake unto me, saying: Thou shalt construct a ship, after the manner which I shall show thee, that I may carry thy people across these waters."[57]

---

[56] Ether 2:16-18
[57] 1 Nephi 17:8

The Angel of Death and the First Passover, Charles Foster, 1897

THE OPEN DOOR

Noah was safe in the ark and all who had responded to his invitations and warnings were also there with him. His wife, his three sons and their wives were all there too. This was the day Noah had long spoken to them about. All the animals, clean and unclean, the reptiles and insects were all there too, content in the places that had been assigned to them. Everything was done as God had said, and so God acted, and He shut the door[58].

God had often revealed Himself to Noah and was with him now. But he didn't say, "Noah, the time has come for us to float away. Close the door." He didn't tell Noah to give orders to the boys to close the door. Their safety in the flood was not going to hang

---

[58] Genesis 7:16

on themselves. The Lord himself closed the door. He shut what "no one can shut, and what he shuts no one can open"[59]. God sealed the door and ensured the complete security of all He had told to enter the Ark. The women wouldn't wake up the first night on the Ark and nudge their husbands and say, "Did you remember to shut the door?" All eight persons on board knew that the Lord had done it. They could rest in peace month after month. They were all perfectly safe. The Lord who had come to Noah all those many years ago and told him to begin this work had Himself completed it by closing the door.

The Lord was the architect of the Ark. He gave Noah specific instructions. Like the other boats built at the Lord's direction; Nephi's Ship, Moses' Basket, and the Jaredite's Boats, God provided the details. In Genesis 6:16 the Lord told Noah, "Make a roof for the ark, and finish it to a cubit above, and *set the door of the ark in its side*. Make it with lower, second, and third decks."

One chapter later we learn, "And those that entered, male and female of all flesh, went in as God had commanded him. *And the Lord shut him in.*"

The Hebrew term used here (translated as "shut him in") is "cagar" (וַיִּסְגֹּר). This is not the first time this word is used in the

---

[59] Revelations 3:7

Book of Genesis. It first appears in Genesis 2:21, but is there translated as "and closed up":

> "And the Lord God caused a deep sleep to fall upon Adam, and he slept: and he took one of his ribs, *and closed up* the flesh instead thereof; And the rib, which the Lord God had taken from man, made he a woman, and brought her unto the man."[60]

Of the seven days that Noah was in the Ark prior to the flood, the Jewish Midrash says about the death of Noah's grandfather Methuselah: "At his death, a frightful thunder was heard, and all beasts burst into tears. He was mourned 7 days by men, and therefore the outbreak of the Flood was postponed till the mourning was over."[61]

There are also Jewish traditions that refer to this time period in astronomical terms: "(in the week of mourning for Methuselah, God caused the primordial light to shine)..." and refers again to the timing of the death of Methuselah, "...God did not wish Methuselah to die at the same time as the sinners..."[62]

---

[60] Genesis 2:21-22
[61] Midrash folio 12
[62] Ginzburg, *Legends of the Jews*, Volume V, 175

The Talmud says there were "flashes of lightning and thunder a very loud sound was heard in the entire world, never heard before" and during those short days the "Holy One...reversed the order of nature, the sun rising in the west and setting in the east."[63]

But after this seven days of morning for Methuselah, it was time. The Lord has an affinity for open doors. He spoke of them directly in Matthew 7:7-8; "Ask, and it shall be given you; seek, and ye shall find; knock, and it shall be opened unto you: For every one that asketh receiveth; and he that seeketh findeth; and to him that knocketh it shall be opened." But now it was time to close a door, to move on.

We find the theme of the closed door of protection running through the scriptures. In Sodom, Lot was standing in the street trying to speak sense into the ears of a wicked group of men, when God's messengers "put forth their hand, and pulled Lot into the house to them, and shut the door."[64] Lot was safe.

In Egypt the angel of death didn't push open the door of the Israelites and take the lives of their firstborn sons when he saw the blood of the lamb on the door frames of their houses.[65] The

---

[63] The Book of the Righteous & Babylonian Talmud
[64] Genesis 19:10
[65] Exodus 12

door was secure; death did not enter their homes; they were safe.

In Jericho, Rahab and her family went into the house, the door was closed and while the rest of Jericho was destroyed, they were safe.[66]

In Jerusalem the prophet Isaiah was God's spokesman addressing the remnant of Judah who were taking refuge from the Assyrians as they invaded the land, "Come, my people, enter thou into thy chambers, and shut thy doors about thee: hide thyself as it were for a little moment, until the indignation be overpast."[67]. They were safe.

In Matthew, Chapter 25, Christ told the story of ten women on their way to a wedding, and half of them were very foolish because they failed to bring any oil for their lamps. They sat around waiting hours for the bridegroom to arrive and finally went to sleep. Then suddenly he was there. The bridegroom had arrived. Half of the welcoming women ran out into the street and lit their lamps of greeting. The other five were looking for oil and missed the feast.

The bridegroom brought the five virgins into the feast. He brought them into his home, "The virgins who were ready went

---

[66] Joshua 2,6
[67] Isaiah 26:20

in with him to the wedding banquet. And the door was shut."[68] The feasting began with all the joy of being with the bride and her groom. As time went by they could all hear knocking on the door from out in the darkness. They could hear shouts. "'Lord, Lord,' they said, 'open the door for us!'"[69] But the opportunity for opening the door had gone by. The bridegroom "...replied, 'Truly I tell you, I don't know you.'" The door to the wedding feast was forever closed to those foolish women. They made no preparation for the greatest event of the year.

The men and women of Noah's day were being confronted with the greatest natural event in history. They had seen the Ark being constructed for decades. But they had never seen the door of the Ark closed. It had always been open as Noah's sons went back and forth carrying tools and wood, and their wives coming and going with food and drink, and Noah preaching to the curious, and the mockers in front of the open door. That was what they thought of when people asked them if they had seen the Ark.

But then, people could see that the door was shut. And it had begun to rain; the skies were black with clouds, and the lightning flashed and then came a deluge like they'd never seen before. The rain came down and seemed to never stop. The

---

[68] Matthew 25:10
[69] Matthew 25:11

rivers quickly overflowed their banks, and the waters rose and filled their houses and swept them away. And then the water level crept up the hills, up and up, never stopping day or night, never slacking, down it came. But the door was shut.

Figure 5: Noah's Door (the Ark Encounter, Kentucky)

In these latter days, the time will come when our protection will be provided by closed doors. While it will not be flood waters, the flood of pornography, same-sex marriage, child abuse, and other carnal enemies will pound at our door. As Noah, we must be prepared when the Lord Himself shuts the door. As Noah, we need to prepare. The Lord told Noah, "And take thou unto thee of all food that is eaten, and thou shalt gather it to thee;

and it shall be for food for thee, and for them."[70] Now is the time to enjoy the open door, but prepare for the day when protection will come through the closed door.

---

[70] Genesis 6:21

THE FOUNTAINS OF THE GREAT DEEP

"In the six hundredth year of Noah's life, in the second month, the seventeenth day of the month, the same day were all the *fountains of the great deep broken up*, and the windows of heaven were opened."[71]

The windows of heaven, and the rain that fell, will be discussed elsewhere, but here the focus will be on the "fountains of the great deep". In a June 2014 article in the journal Science[72], it was reported that massive amounts of water appear to exist deep beneath the planet's surface, trapped in a rocky layer of the mantle at depths between 250 and 410 miles. "It may equal or perhaps be larger than the amount of water in the oceans,"

---

[71] Genesis 7:11
[72] Science, 13 June 2014, V344, I6189, 1265-1268

Northwestern University geophysicist Steve Jacobsen said. He continued, "It alters our thoughts about the composition of the Earth."

During the flood, God opened the fountains of the deep. When God ended the flooding, we are told that God closed the fountains of the deep, suggesting that they continue to exist. And now we have confirmation that they do.

The water does not exist in huge underground oceans. The water is tied up in minerals, held there by the tremendous pressures and temperatures deep in the earth. Researchers have analyzed the amount of this water that is brought to the surface during volcanic eruptions and concluded from their findings that there could be enough water deep in the earth to fill many, many oceans. This water is stored in a very unique manner, allowing it to be stored up, and then released only in particular circumstances.

> "When he prepared the heavens, I was there: when he set a compass upon the face of the depth: when he established the clouds above: when he strengthened the fountains of the deep."[73]

When the Lord released these fountains, as we saw with accounts of Methuselah's death previously, there were reports

---

[73] Proverbs 8:27-28

THE FOUNTAINS OF THE GREAT DEEP

of great thundering, and cracking. As the earth began to break apart, Rodinia, the proto-continent was split into pieces or plates. As it did, tremendous forces were at work, releasing this water that was held in reserve, deep in the earth.[74]

The result was not just rain from the sky, which resulted from the oceans heating, and clouds being filled to saturation, then dumping huge quantities of water, but these fountains also spewed great geysers of water, adding even more to the floods.

Because Rodinia was largely a flat land, the water trapped in the earth, and the waves created by huge sections of land splitting apart and moving, quickly inundated the land. Throwing waves of water, full of sediment, animals and plants deep across the continent. Wave after wave, ecosystem after ecosystem, scrubbed from the depths of the sea, and the plains and hills of the earth.

To create some context for the effect of these waves, consider that if all the current land masses were flattened and spread equally across the surface of the planet, water would cover the entire surface to a depth of almost 9,000 feet (approximately 8,858 feet to be exact). So, while so much discussion of the reality of the flood focuses on the challenges of a water depth

---

[74] Andrew A. Snelling, Geological Issues: Charting a scheme for correlating the rock layers with the Biblical Record, *Grappling With the Chronology of the Genesis Flood*, Master Books, 77-109

sufficient to cover the highest peaks (over 29,000 feet for Mount Everest), in reality this is not a challenge. In fact, prior to the opening of the fountains of the deep, there may even have been less water on the surface, but with the additional water, somewhere akin to what we have now, there was definitely sufficient to cover all land.

These sequential waves deposited their debris in layer after layer, creating the great sequences of sediments so visible in the Grand Canyon and elsewhere. In some cases, the circumstances were favorable to fossil formation, in others they were not, but in each case, the ecosystems were ravaged by ocean waters, and the debris they carried.

There are still remnants of this early protocontinent, Rodinia in the deep bedrock, known as cratons. These cratons allow us to put together the pieces of the early continent. They range from what is now North America to present day Scandanavia, Africa, India and Eastern Europe. This is the continent before the sediments extended it size, and before the collisions to come.

Keith James, a geologist from the Institute of Geography and Earth Sciences of Aberystwyth University in Wales, UK, stated that geoscientists still do not understand the origin of continents. Continental crust is supposed to be produced "by complex partial melting of sediments, the [subducted] slab, the mantle or the mantle wedge (or combinations of these) in

THE FOUNTAINS OF THE GREAT DEEP

'subduction factories.'" However, geoscientists readily admit the continents are too large for this simple explanation.[75]

As Rodinia split, the ocean crust, which was a precarious heavier layer riding on a lighter basalt, began to sink into great crevasses. These are called subduction zones, and they exist today where these plates continue to collide. In normal circumstances, with today's forces, this movement is slow. Just inches per year. But, when additional forces are at work, and when it is driven by the difference in density of these two layers (as it was leading up the flood), it can begin to move more rapidly. As it does, it creates greater friction, which melts the rock, creating a lubricant, which further increases the subduction, which adds more friction, and so on. It is estimated that rather than the inches per year that are currently seen, this could result in feet per minute. This would have the effect of pulling Rodinia apart as the ocean crusts dove beneath the surface, and new crust rose to the top. This would further change the pressures deep below the surface, causing even more water to be released from these great fountains of the deep.[76]

---

[75] Keith James, 2018, Not Written in Stone: Plate tectonics at 50, *AAPG Explorer*. 39(2):18-23
[76] Jeff Hecth, 2015, Rise of the upper crust. *New Scientist,* 226 (3017): 36-39

"In the South Atlantic, granite [continental rock] was in 2013 discovered on the northwest-to-southeast Rio Grande Ridge (outer edge of South America's magnetic extension)." Standard tectonic theory cannot explain the presence of these continental rocks and fossils found so deep in the ocean basins.[77]

The initial result was the movement of these pieces of Rodinia into a pattern that created Pangea. This continent would have existed below the waves of Noah's flood, and would have existed for just a short time, perhaps just weeks, before it broke apart again as the forces in the earth's crust continued to work.

---

[77] Keith James, 2018, Not Written in Stone: Plate tectonics at 50, *AAPG Explorer,* 39(2):18-23

THE FOUNTAINS OF THE GREAT DEEP

Figure 6: Pangea (The Submerged Continent)

It is worth noting that those who tried to utilize Pangea as the protocontinent, the place where Adam and the generations to Noah would have lived, have always faced challenges, due to the existence of the cratons that predate it. The introduction of Rodinia fits all the pieces into place, and results in a fossil record that supports the scriptural account. Pangea helps us understand the current continents, but Rodinia helps us to understand the world of creation.

We can be certain of their existence, and the basic timeline, due to the sediment formation we observe. Early layers represent an early deposition of sediments, that then experienced some outside force (the collision of Rodinia fragments into a Pangea intermediate continent) to cause them to bend and buckle, shortly after their creation. These folded sediments, involving multiple distinct layers, can be seen throughout the world. But, Pangea was not to remain, and split again into plates that continued to move, with more sediments being deposited as the flood waters continued to move over the earth.

These sections of crust collided again, resulting in the continents we see today, and creating the massive mountain ranges we observe. The marine fossils that exist at the top of some of the high peaks are a testament to both the flood waters that accompanied their initial creation, as well as the movement of plates that provided the massive amounts of water needed. For years these marine organism fragments have been the source of questions, now they serve as further evidence of the reality of a universal flood. The very earth beneath our feet is a testimony to the truthfulness of God's word.

THE RAVEN AND THE DOVE

After 40 days passed, Noah opened the hatch of the ark and sent out a raven. The raven is a scavenger that feeds off the flesh of the dead. However, the raven did not return. So, Noah let seven days pass and then sent out a dove that soon returned. Noah let another seven days pass and sent out the dove a second time. It returned again, but this time with an olive tree leaf in its beak. Noah sent out the dove a third time and this time the dove did not return. The time to leave the ark and begin a new life on earth had arrived.

There is some interesting symbolism expressed through the raven and the dove.

In Jewish tradition, the raven is considered an unclean bird, and is considered to be a symbol of evil. In contrast, the dove is a clean bird and in scripture is a symbol of the Holy Spirit.

Notice the characteristics portrayed of the raven. The Scripture says that the raven went forth "to and fro" and never returned. Who else in Scripture is described as roaming the earth to and fro? None other than Satan himself. Second, the raven is a bird that feeds on the flesh of the dead.

Now, whereas the raven was only sent out once, the dove was sent out three times. The number three holds a special significance in the Scriptures. For example, Noah had three sons, Christ was raised on the third day, and the ark of the covenant contained three sacred objects and there are many other examples.

The harbinger of Noah's exit from the ark must come not from an animal that harkens back to the sins of the past, and that literally feeds on the destruction they caused, but from one that helps Noah and his family begin afresh, looking for new growth, as the dove did.

In comparing the raven with the dove, it should be noted that while the raven was able to derive satisfaction from the dead fleshy things of the world, the dove wasn't able to and thus returned. When the dove came back a second time, it had an olive tree leaf in its beak. When it finally did not return, it had built a new home.

The olive tree is seen as a symbol of peace. But, in addition, olives themselves have been known throughout the ages for their nutritional and healing value. In the Bible, olive oil has played a significant role in anointing kings and high-priests and was used to light the menorah in the Holy Temple.

Using these two birds in particular is great reinforcement for the events which just took place in the world. In one case, a dark, carnal bird, one that eats dead flesh and is seen widely as a symbol for evil. In the other, a white, clean bird, who requires a safe, secure nesting place and is seen as a symbol of peace and the Holy Spirit.

The raven would be a sign that the world was still a place of death. Only the dove would be a suitable sign that the world was ready to enter into its new creation.

THE FOSSIL RECORD

Perhaps some of the most compelling evidence for the existence of the flood lies right beneath our feet. While fossils have been a stumbling block to many who want to believe the scriptural account, it has been difficult to understand how they fit. The reality is they are a window into the events of the flood. They make it real, they testify of the reality of this incredible event.

Imagine a leaf, falling from a tree. Autumn winds tug at it until finally it separates from the tree which just months ago counted on it to gather the energy of the summer sun. But now winter is coming, the days are growing shorter, and the temperatures cooler. Soon it will not be a benefit, and the snow and ice it will capture will endanger the tree. So, the tree sheds these unnecessary parts. They fall to the ground, and are blown with the wind.

Have you noticed what happens to these fallen leaves? They shrivel and twist themselves, quickly becoming brittle and then breaking into fragments. If you want to capture the beauty of an autumn leaf, you have to get it quickly, just after (or preferably just before) it falls. We all did this as children. Grabbing these beautiful leaves, and then pressing them between the pages of a heavy book. In this way, we were able to protect them, and keep them flat and beautiful.

When an animal or insect dies, the process is similar, but much more violent. Each animal that dies, even if from old age, quickly becomes a food source for other animals. It will be devoured first by scavengers, who pull the flesh from the bones, and then by organisms that decay even the smallest fragments. Rarely will we observe a complete skeleton of a dead animal, its bones will have been scattered over a distance as predators and scavengers remove every bit of food they can from the remains. If a hunter or museum wants to display a specimen, they gather it quickly, preserve it, and protect it to ensure pieces are not damaged or lost to natural processes.

This is the world we observe. These are the forces we understand.

It is a surprise then, that fossils exist at all. The circumstances necessary for their creation seem to indicate that they are unlikely. How do we have fossils of perfectly flat leaves, when this is such an apparently unnatural circumstance? And how could there possibly be fossils of animals when scavengers would have scattered the bones? There is only one logical set of circumstances that can explain their existence.

Just as the pressure of a books pages can prevent a leaf from shriveling, the force of many feet of floodwater mud can accomplish the same feat. The same can be said for the animal fossils that have been discovered. Only an event that would surround these carcasses completely and quickly would have the ability to allow the processes necessary for fossil formation to occur, and it would have to be immediately after (or in fact the actual cause of) death.

A little background on what a fossil actually is may be helpful here. Fossils can form in a wide variety of ways. Some common methods include:

- The body can leave an impression or mold showing its outer shape in the surrounding sand or mud. This can include footprints and the inside and outside of shells. These fossils typically do not display cellular structure, but will look like a mold was filled with material that then became stone.

- Petrification takes place when minerals replace the original material of the plant or animal. Petrified fossils must form quickly, before the body parts have time to decay. Petrified wood is a classic example. These will often show anatomical structures. Bones and other physical elements will be visible, caused by the decay of body parts at different rates, and the replacement by different minerals, creating a stone version of the original organism.

Figure 7: Petrified Wood

- Permineralization, or encased fossilization, occurs when dissolved minerals fill the pores and empty spaces in the plant or animal but don't replace any of the original material. The chemicals then turn into crystals, keeping the organism safe and preserved. While it is possible for many different chemicals to do this, quartz is the most common. Most dinosaur bones are permineralized.

Fossils can form under all kinds of conditions. Water and dissolved minerals are usually needed to form the three types of fossils above. But, many processes—coalification, compression, freezing, desiccation (drying out), to name a few, do not require water or minerals.

There are those who choose to believe, and even tout as settled science, that this process requires millions, or even billions of years, and is the result of sequential flood events, separated by thousands, millions or billions of years. They claim that these animals simply fell into a body of water, settled to the bottom and were covered by sediments. There are many reasons this interpretation is inaccurate:

- Between each of the discrete layers of sediments, and the fossils they contain, there is little erosion. If these layers had been deposited by sporadic floods, exposed to the elements, then another flood event, we would expect to see erosion features between the layers of sediment. They are practically non-existent. Indicating a rapid succession of layers deposited in a relatively short period of time.

- There are numerous instances of fossils that cross multiple layers of sediment. These are called *polystratic* fossils. These can include shells, trees, and other organisms. There is no logical explanation for their existence that does not involve rapid deposition of multiple layers of sediment over a short period of time.

Figure 8: Polystratic Fossil

- There is a term used by geologists, *paraconformity*. In short, these are fossils of organisms that do not appear in the "proper" place. The view that the organisms fossilized in these sediments represent millions of years of history, and a record of an evolutionary process requires that less complex organisms be at the bottom, and more complex organisms be nearer the top. But, this is often not the case. Paraconformity refers to fossils that do not appear where they should. They may bridge subsequent layers, appearing "millions of years" out of place. The fact that they lived at the same time, were deposited in a global flood, and that flood surges brought organisms in at different rates, does not require a "paraconformity" paradox.

- Soft bodied organisms, such as jelly-fish, should not create fossils if they were deposited in lake or sea sediments, and then covered over a long duration of time. They decompose too quickly. A jellyfish that washes up on the beach is quickly reduced to a pile of slimey, mushy, goo. It does not resemble a jelly fish for long. So, even without taking into consideration scavengers, this organism with no hard body parts would not be expected to result in a fossil, and yet they exist, often with incredible detail. This would require a very short time between their death and their entombment in sediment. In fact the most likely sequence of events would indicate that the sediments were the cause of death.

Figure 9: Jellyfish Fossils (*Heimalora stellaris*)

- Animal track fossils are a remarkable site. They indicate, in many cases, large numbers of animals apparently moving in a common direction. As if they are fleeing some event. The circumstances required to capture a footprint in mud are very unique. We have all stepped in the mud, or in the sand on the beach. How long do these footprints remain? Not long. Like the forensic detectives on our favorite crime drama, a cast must be made quickly, or the integrity is lost. And yet, thousands and thousands of footprints have been preserved, in remarkable detail. The event that captured them must have been unique. A flood of sediments, laid down quickly, and shortly after the animal was present, or perhaps the flood was the reason the animals were fleeing.

Figure 10: Track Fossils

- The broad reach of similar layers of sediment is beautiful to observe. The Grand Canyon provides an amazing perspective on these layers.

THE FOSSIL RECORD   81

Figure 11: Grand Canyon

What is not immediately obvious is the fact that these same layers exist over a much broader reach. The fossil-bearing sediments that make up the Grand Canyon extend for miles, some continent-wide, and even further. They represent a massive deposition of sediments, over a short period of time. As you observe these layers, look for erosion characteristics. They do not exist. There are relatively straight, clear lines that indicate where one kind of sediment was deposited and the next arrived to cover it. Sometimes there will be waves in the sediments, indicating they were bent while still flexible. Some of these bends even span multiple layers, indicating sediments, supposedly deposited thousands of years apart, were all still soft and pliable at the same time.

The reality is that the layers that make up the Grand Canyon, and sediment layers all over the world, were deposited quickly, and in rapid succession. One layer, known as the Basal Sauk Megasequence, covers much of North American (and even into South America).

Figure 12: Basal Sauk Megasequence

- Fossils often capture a very dramatic, and short-lived event. One such type of fossil, which is far from rare, captures one animal in the process of eating another.

Some of my favorites involve fish, and show that the flood came quickly, and serves as a snap shot into a violent event that intruded on otherwise routine circumstances.

Figure 13: Fish Buried in the Act of Eating Another Fish

Discussions of fossils inevitably result in discussions of evolution. Evolution is a challenging topic, because there really is no common reference point for it. Many people who say they believe in evolution, will use as examples, illustrations of natural selection. A very different process as you will see. So, let's first clarify the use of these terms, for the purposes of this book.

I do not believe that evolution occurs, because I apply a definition that requires evolution to describe what others may call macroevolution. Macroevolution is the current term being used to describe the process that ultimately allows a single-cell organism to become you and I. This kind of evolution has no demonstrable basis.

What does occur is what some call microevolution, but what is most often referred to as "natural selection." This is the process where a specific trait may be encouraged in a species, through pressure from the environment, which may ultimately result in a new species. The process is the result of information already present in the DNA of the original organism, which has now been expressed.

In January of 2018, the National Aeronautics and Space Administration (NASA) release a 2018 Investigator's Workshop for NASA's Human Research Program. Scott Kelly is a twin. He spent 340 days aboard the International Space Station (ISS). Because he is a twin, this experience provided a unique opportunity for NASA to track changes in DNA associated with his time in space.

Figure 14: Scott and Mark Kelly

To track physical changes caused by time in space, scientists measured Kelly's metabolites (necessary for maintaining life), cytokines (secreted by immune system cells) and proteins (workhorses within each cell) before, during and after his mission. The researchers learned that spaceflight is associated with oxygen-deprivation stress, increased inflammation and dramatic nutrient shifts that affect gene expression. The environmental stresses caused changes in Kelly's DNA, not due to damage, but directly in response to these environmental conditions.

In particular, Chris Mason of Weill Cornell Medicine reported on the activation of Kelly's "space genes" while confirming the results of his separate NASA study, published last year. The team found DNA changes which resulted in modifications to Kelly's collagen, blood clotting and bone formation due, most likely, to fluid shifts and zero gravity. The researchers discovered hyperactive immune activity as well, thought to be the result of his radically different environment: space.

It is easy to see how these same characteristics would be used as organisms aboard the ark were subjected to a changed, and still changing, world upon leaving the ark. Changes would be necessary to function in this new post-flood world. This is the power of DNA. God's ingenius engineering allows his creations to use the information in their DNA to modify an organism, due to outside circumstances.

To put it simply, the information that allows a new breed of dog, is not new information. The information needed was already present in the DNA of the original breeding pair. What selective breeding does (just a controlled natural selection) is encourage the expression of specific portions of the DNA.

You may not realize that one basic assumption made by Charles Darwin was that the earth was millions of years old, and that the earth has undergone the same physical processes in the past that we observe today. In fact, Darwin, in the first edition of his Origin of Species, estimated that it took 300 million years to erode the Weald, a chalk deposit in southern England.[78] This principle is called uniformitarianism, and is in direct contradiction to scripture, and fulfills prophecy of the last days:

> "...*there shall come in the last days scoffers*, walking after their own lusts, And saying, Where is the promise of his coming? for since the fathers fell asleep, *all things continue as they were from the beginning of the creation.*"[79]

Darwin's opinions were far from unanimous, even in the nineteenth century. None other than Lord Kelvin, of the Glasgow University, and widely regarded as the best physicist of

---

[78] Anthony Hallam, *Great geological controversies*, Oxford University Press, New York, 106; Cherry Lewis, *The dating game: One man's Search for the age of the Earth*, Cambridge University Press, Cambridge, United Kingdom, 25
[79] II Peter 3:3-4

THE FOSSIL RECORD

his time, (his expertise on thermodynamics was particularly revered) felt that the earth was much younger. His calculations, which did not benefit from modern knowledge, continuously resulted in younger and younger ages for our planet. Even the noted author Mark Twain weighed in on the debate, he stated:

> "Some of the great scientists, carefully ciphering the evidences furnished by geology, have arrived at the conviction that our world is prodigiously old, and they may be right but Lord Kelvin is not of their opinion. He takes the cautious, conservative view, in order to be on the safe side, and feels sure it is not so old as they think. As Lord Kelvin is the highest authority in science now living, I think we must yield to him and accept his views."[80]

There is no valid reason to expect that the same processes at work today have continued unabated since creation. Evidence used to support geological dating is flawed, and employs circular reasoning. Beginning with the premise that fossils are millions (or billions) of years old, and assuming sediment deposition happens at the same rate now as it did then (ignoring an event such as a global flood), they create a presumed age for fossils. They then use the presence of these fossils elsewhere to support the age of new sediment deposits.

---

[80] Mark Twain, *Letters from the Earth*

They take another step into this flawed world by using what appears to be a logical assumption, that these layers of sediments indicate a chronology of evolution. That each subsequent layer up represents not only millions of years of sediment deposits, but millions of years of evolution. They portray the fossils these organisms represent, as snapshots in an evolutionary movie, with less complex organisms at the bottom, and humans at the top. Never mind that the lower organisms are immensely complex, or that organisms often jump from one layer to another, or that the order is often rearranged. In the never-ending effort to justify the evolution theory, and to dismiss God and his creation, they ignore real science.

"But ask now the beasts, and they shall teach thee; and the fowls of the air, and they shall tell thee: Or speak to the earth, and it shall teach thee: and the fishes of the sea shall declare unto thee."[81] The answers are available. In the rocks, the records of the beasts remain.

A few examples may be beneficial. First let me discuss what is often portrayed as a very early organism, low in the column of sediments, and therefore not very complex in biology.

---

[81] Job 12: 7-8

THE FOSSIL RECORD

Figure 15: Trilobite

Because trilobites occur in such large numbers, and in this lower sediment deposit, it is often assumed they are the oldest representatives. This requires they be less complex, although in the deposits where trilobite fossils are found, there is already incredible diversity. The variety of body types is amazing; horns, antennae, shapes in its complexity. There are species with no eyes, and species with complex eyes. So, rather than being a record of an evolutionary process, they are perfect examples of God's command to "Let the waters bring forth abundantly the moving creature that hath life..."[82]

---

[82] Moses 2:20

Figure 16: Trilobite Variation

They lived in a marine environment, so it should come as no surprise that they are the first animals to be covered by the sediments created by the flood.

Another common claim is that they are extinct, which evolutionists somehow think indicates that they evolved into something else. But, just considering the biology of the trilobite's eyes, there is a living member, the horseshoe crab, Limulus. Limulus is often called a "living fossil." This is a term used by evolutionists to describe an organism that is represented in both the fossil record, and the living world. The world is full of such "living fossils."

Figure 17: Horseshoe Crab, living example

Figure 18: Horseshoe Crab, fossil

Those who choose to embrace evolution and the "accidental" creation of life, require millions of years, leaps in chemistry and biology, cataclysmic asteroids to produce extinction events, and a disdain for God, to fuel their belief in evolution.

A recent discovery indicates the extent those who embrace evolution will go to. Hesham Sallam of Mansoura University in

Egypt revealed in February, 2018 that they had found fossils left by a sauropod (a large, long-necked dinosaur) in Africa.[83] It may come as a surprise that no fossils for such a dinosaur had been found on the African continent prior to this. In fact, another member of the team, Marr Lamanna of Carnegie Museum of Natural History, Pittsburgh stated "When I first saw the pics of the fossils, by jaw hit the floor. This was the Holy Grail...that we paleontologists had been looking for a long, long time." Their explanation for this animal's existence in Africa? Because of its relationship to similar dinosaurs in Eurasia and Africa, apparently it swam there. A large, long-necked dinosaur swimming to Africa. They would rather believe that then that a global flood deposited it there.

While many had believed that the days of trying to turn lead into gold were past, the idea of an organism expressing information that its DNA never included (ie a slug becoming a bear) is nothing more than genetic alchemy. It may be wished for, but it is not a reality. It is unfortunate that evolution is presented as settled science, when in reality it is not. It is based on one flawed assumption after another. It has as a premise that the earth is millions of years old, that in spite of all evidence to the contrary nature has gone from less complex to

---

[83] Hesham M. Sallam, "New Egyptian Sauropod Reveals Late Cretaceous Dinosaur Dispersal Between Europe and Africa", *Nature Ecology & Evolution 2*, 445-451, 29 January 2018

more complex, and that the incredible, adaptive, flexible engineering required to embed information in the DNA in a way that allows the needed code to be available for these modifications is all a function of random events.

Figure 19: Sauropod Dinosaur

It is only a matter of time until DNA is extracted from these fossils. While evolutionary science denies this possibility, due to the millions of years they claim for the age of these fossils, the reality is that biological evidence has been found, which clearly invalidates the claimed age of these fossils.

There have already been dozens of examples of collagen, skin pigments and other biological artifacts remaining in these partially fossilized organisms, some with claimed ages of tens of millions of years. Even though these materials typically decay within hundreds or thousands of years. One important recent discovery found fat from fossilized bird glands, the kind birds use to preen their feathers to make them water-proof. These glands still containing original fat were called "the oldest lipids ever recovered from a vertebrate."[84] Lipid describes fats and oils. The scientists describing this find claimed it was 48 million years old. Even though there is no accurate way to determine age, and clearly these fats would have been attacked by bacteria long ago. It is actually surprising that they would survive even the thousands of years they did.

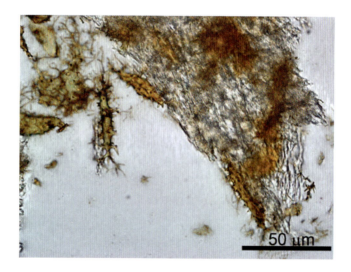

Figure 20: Red blood cells in a *hadrosaur* fossil. (Mary H. Schweitzer)

---

[84] Micu, A., Researchers discover astonishingly intact, 48-million-year-old bird fats in fossilized gland. *ZME Science*, October 19, 2017

THE FOSSIL RECORD

Carbon-14 dating is often used to justify the age of organic material. It is problematic since "that assumes that the amount of carbon-14 in the atmosphere was constant — any variation would speed up or slow down the clock."[85] If we abandon uniformitarianism (which we should), we understand that there is no reason to expect that carbon-14 levels would have remained constant. The world was a very different place prior to the flood, and many processes would seem very foreign to us.

In an embarrassing moment, living freshwater claims were radiocarbon dated to be in excess of 1600 years old[86]. Samples of coal confuse scientists, who want to rely on radiocarbon dating, but then refuse to believe their own results which state that coal must be less than 20,000 years old (when they claim it is millions of years old)[87].

In spite of the challenges, scientists insist on using the questionable data to support their theories. "Despite its overuse

---

[85] Ewen Callaway, Carbon Dating Gets a Reset, *Scientific American*, October 18, 2012
[86] Taylor, R. E., *Radiocarbon Dating: An Archaeological Perspective*, 1987
[87] Bowman, Sheridan, 1990, *Radiocarbon Dating*, Berkeley: University of California Press

and misrepresentation in the media, it is nonetheless extremely valuable,"[88] says Michigan State University.

Unfortunately, at times they go even further, with obvious efforts to twist evidence to support their flawed theory. None other than the prestigious National Geographic Society has used their platform to promote this false science. In the November 1999 issue of their magazine, National Geographic, an article was published titled "Feathers for T. Rex?"[89], that purported to demonstrate "a true missing link in the complex chain that connects dinosaurs to birds." What was not known until later was, not only was this fossil a fraud, but questions raised prior to the publication, by none other than Dr. Storrs L. Olson, curator of birds at the Smithsonian Institution's National Museum of Natural History were ignored. He wrote an open letter accusing National Geographic of "an all-time low for engaging in sensationalistic, unsubstantiated, tabloid journalism."[90]

---

[88] Andrea Cohn, Radiocarbon Dating: A Closer Look At Its Main Flaws, *Great Discoveries in Archaeology*, Spring 2013, Michigan State University
[89] Christopher P. Sloan, Feathers for T. Rex?, *National Geographic Magazine*, 1999, 196[5], 100
[90] Dr. Storrs L. Olson, "open letter to Peter Raven, Secretary of the National Geographic Society Committee for Research and Exploration", November 1, 1999

THE FOSSIL RECORD

Figure 21: November 1999 National Geographic

When he was invited to view the evidence being used by National Geographic to support its article, prior to publication of the article, he stated that it "became clear to me that National Geographic was not interested in anything other than the prevailing dogma that birds evolved from dinosaurs."

Speaking further on the subject he wrote:

> "The hype about feathered dinosaurs in the exhibit currently on display at the National Geographic Society is even worse, and makes the spurious claim that there is strong evidence that a wide variety of carnivorous dinosaurs had feathers. A model of the undisputed dinosaur Deinonychus and illustrations of baby tyrannosaurs are shown clad in feathers, all of which is

simply imaginary and has no place outside of science fiction."[91]

His conclusion is very telling, and relates not only to this incident, but to the efforts by evolutionists in general:

"The idea of feathered dinosaurs and the theropod origin of birds is being actively promulgated by a cadre of zealous scientists acting in concert with certain editors at Nature and National Geographic who themselves have become outspoken and highly biased proselytizers of the faith. Truth and careful scientific weighing of evidence have been among the first casualties in their program..."[92]

As you can see from his comments, this has become a false religion, an object of faith, with truth and science being a casualty of their evangelism. "... When they are learned they think they are wise ... supposing they know of themselves, wherefore, their wisdom is foolishness. ... And they shall perish."[93]

---

[91] Olson, 1999
[92] Olson, 1999
[93] 2 Nephi 9:28

Thankfully, modern prophets continue to speak eternal truths:

> "No lesson is more manifest in nature than that all living things do as the Lord commanded in the Creation. They reproduce 'after their own kind.'[94] They follow the pattern of their parentage. ... A bird will not become an animal nor a fish. A mammal will not beget reptiles, nor 'do men gather ... figs of thistles'"[95],[96]

---

[94] Moses 2:12, 24
[95] Matthew 7:16
[96] Boyd K. Packer, Conference Report, October 1984, 83

100 THE TOKEN OF THE BOW

"*Rainbow Dragon,*" Tom Kyle

HERE BE DRAGONS

On one of the world's oldest surviving globes, the Hunt-Lenox Globe (from about 1510), near the edge of the known world, are the words "HC SVNT DRACONES". In Latin, hic sunt dracones is translated as "here be dragons".

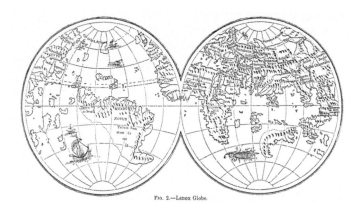

Fig. 2.—Lenox Globe.

Although this is the only surviving map or globe bearing this specific phrase, there are several more cases where cartographers implied fantastic dangers existed beyond the boundaries of their knowledge, with pictures of beasts or phrases such as hic abundant leones ("here lions abound"). Even Ptolemy's atlas in Geographia, a compilation of the geographical knowledge of the Roman Empire in the 2nd century, warns of elephants, hippos and cannibals lingering in unknown lands.

While hardly the only "living fossil" we have knowledge of, nothing creates quiet the excitement of a good dragon story. From the Archangel Michael and Saint George slaying dragons, to ancient Chinese artwork, dragons occupy our imaginations. But, there is perhaps more to the story that simply a child's fairytale, or some artists wild imagination.

If I were to describe to you an animal that was massive in size, was carnivorous, aggressive, had a piercing roar, and who's skin seemed to resemble a knight's armor, what would you think? Our current perceptions of a dragon might describe it well. We might even describe a range of creatures with this title. It could even come to describe any large, frightening, scaled beast.

Figure 22: "Dragon Flight" (Jackson Obrien, 2018)

The real question is, whether there is any reason to believe that such a creature ever existed alongside people. If we believe the scriptures, it must.

The word dinosaur wasn't even around until scientist Sir Richard Owen introduced it in the mid-1800s. Before then, large reptiles were called dragons. In fact, dictionaries in the

1600's defined dragons as "rarely seen but still living creatures."

The 19th century English clergyman Charles Kingsley observed, "Did not learned men, too, hold, till within the last twenty-five years, that a flying dragon was an impossible monster? And do we not now know that there are hundreds of them found as fossils up and down the world? People call them Pterodactyls: but that is only because they are ashamed to call them flying dragons, after denying so long that flying dragons could exist."[97]

In 326 AD, Alexander the Great invaded India. While on this campaign, he wrote a letter to his long-time teacher, Aristotle, giving an account of his interaction, and subsequent killing of a dragon:

> "And on the second night, behold, [there came] a monster a beast which was larger than an elephant, and when it drew nigh to the ditch within which we had entrenched ourselves it wished to escape, but it was not able to do so. Now when I saw this, I commanded that herald to go round about through the camp [and to order the soldiers] to put on their armour and to protect themselves that night, and I informed them concerning

---

[97] Charles Kingsley, *The Water Babies*, 1863

the beast the like of which I had never seen before; and I commanded them, moreover, to light fires round about them, and that each man should abide in his own place the whole of that night. And when the beast saw the fighting men it lifted itself up and betook itself to flight, but as it was fleeing, by reason of its mighty rage it fell into the ditch within which we had entrenched ourselves; and I commanded thirty of our mighty warriors to rise up against it and to slay it, and they slew that beast. And it came to pass on the morrow when daylight had appeared that I commanded the men to bind the beast with ropes, and three hundred men dragged it out of the ditch, and cut open its belly, and they found therein great numbers of snakes and scorpions, and fish larger than oxen; and each of its tusks was a cubit in length, and its claws were like unto those of hawks. Now this beast is the most voracious of all the wild beasts of that country."[98]

Dragons feature prominently in English history as well. In fact, the story of King Arthur (his father's name was Uther Pendragon, "dragon's head") is predicted in the story of the Red Dragon and the White Dragon in *History of the Kings of*

---

[98] *Alexander's Epistle to Aristotle*, trans. Budge, 1896, p149-150

Britain[99]. The Red Dragon made its home in what is now Wales, and was invaded by the White Dragon. The flag of Wales features a red dragon today.

The Hebrew word tannin (תַּנִּין) is defined as "serpent, dragon, sea-monster." It likely refers to certain reptiles, including giant marine creatures and serpentine land animals. Though translated several different ways and differing in precise meanings based on context, tannin can denote a dragon and therefore can potentially refer to a dinosaur since all dinosaurs are dragons (though not all dragons are dinosaurs by definition). We see mention of dragons throughout the Old Testament, from the prophets Nehemiah, Isaiah, Ezekiel, Micah and Malachi, as well as in the Book of Mormon with Nephi embracing the words of Isaiah. There are also the statements of John in the Book of Revelations, comparing Satan to a dragon in the last days.

---

[99] Geoffrey of Monmouth, *The history of the kings of Britain: an edition and translation of De gestis Britonum*)

Why isn't tannin translated as "dragon" in some more recent English versions? Perhaps it is due to many misunderstandings about what dragons really were. In other similar instances we find that translations list elephant or hippopotamus in the footnotes in Job 40 when discussing behemoth. Let's look closer at the behemoth to give you some context. In the book of Job, God describes the behemoth that "eats grass like an ox" and "moves his tail like a cedar" with bones that "are like beams of bronze"[100]. The beast the passage describes fits well with something similar to a sauropod dinosaur like Brachiosaurus.

Next, God describes at length a leviathan, a fire-breathing sea monster with impenetrable scales that none could face except its Creator. Read Job 41 and see if you picture a ferocious marine reptile, like a Kronosaurus. Leviathan is mentioned in five passages of Scripture and is identified as a type of tannin in Psalm 74:13–14 and Isaiah 27:1. Dragons are real, created creatures, some of which terrorized in the waters and others that roved the land and air.

---

[100] Job 40:15–18

Figure 23: *Kronosaurus queenslandicus*

The word "dragon" is derived from the Latin, draconen meaning "huge serpent" and the Greek, "drakon" (δράκων) which is defined as a "serpent". In fact the Ethiopian Dragon was a breed of giant serpent native to the lands of Ethiopia. They are said to have killed been so large they killed elephants. They are mentioned in the work of Aelian[101].

Land and air dragons would have been taken on Noah's Ark and probably existed for some time afterward, based on the descriptions we see in the Bible and legends and artifacts worldwide. But they died out due to factors such as climate changes (the post-flood ice age), food source problems, genetic mutations, and diseases. Man most likely played a role in the demise of dragons, as we read in the legends of dragon slayers such as Saint George.

---

[101] Aelian, *On the Characteristics of Animals*, 2.21

The word dinosaur literally means "terrible, powerful, wondrous reptiles." The word became popular in 1841 after biologist and paleontologist Sir Richard Owen coined the term "Dinosauria"[102].

Figure 24: Sir Richard Own

His justification, which focuses mainly on their size:

> "The combination of such characters, some, as the sacral bones, altogether peculiar among reptiles, others borrowed, as it were, from groups now distinct from each other, and all manifested by creatures far surpassing in size the largest of existing reptiles, will, it is presumed, be deemed sufficient ground for

---

[102] Richard Owen, *Journal of the British Association for the Advancement of Science (BAAS)*, 1841

establishing a distinct tribe or suborder of Saurian Reptiles, for which I would propose the name of Dinosauria."[103]

Figure 25: archaeopteryx lithographica

---

[103] Richard Owen, *Journal of the British Association for the Advancement of Science (BAAS)*, 1841

Owen was the first to recognize the great fossilized bird, Archaeopteryx lithographica. This bird, often called a "dinosaur with wings", is actually a full-fledged bird. It contains all the characteristics necessary to be classified as a bird, and as we know from Genesis 1:21, God created "every winged bird according to its kind." Even Charles Darwin agreed, he mentioned it in the fourth edition of *Origin of the Species*, describing it as merely "a strange bird". A recent BBC report spoke of new evidence that leads to the belief that Archaeopteryx flew like a pheasant[104]. X-ray results revealed bones that were almost hollow, just like modern birds[105]. Dennis Voeten, from the European Synchrotron facility in Genoble, France stated that; "We imagine something like pheasants and quails...If they have to fly to evade a predator they will make a very quick ascent, typically followed by a very short horizontal flight and then they make a running escape afterwards." *(For more on dinosaur "birds" see the chapter, "The Fossil Record").*

---

[104] Helen Briggs, Archaeopteryx flew like a pheasant, say scientists, BBC News, 13 March 2018
[105] Dennis F.A.E. Voeten, Wing bone geometry reveals active flight in Archaeopteryz, *Nature Communications* 9, Article: 923, 2018

Figure 26: *archaeopteryx* in Flight (JANA RŮŽIČKOVÁ)

The word dinosaur derives from two greek words, deinos (δεινῶς) "terrible, powerful, wondrous" + (sauros) "reptile or lizard". Prior to 1841, what were people to call these "terrible lizards"? Dragons.

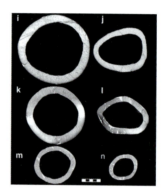

Figure 27: X-rays of archaeopteryx Hollow Bones

In 2004 a unique fossil skull was donated to the Children's Museum of Indianapolis for study. Unique because of its somewhat fanciful name, Dracorex Hogwartsia (which would translate at Dragon King of Hogwarts), and because of its startling similarity to the subject of this chapter...a dragon. This skull and several vertebrae were discovered by Bob Bakker and Robert Sullivan[106].

Figure 28: *Dracorex Hogwartsia*

Harry Potter author, J.K. Rowling had this to say:

"I am absolutely thrilled to think that Hogwarts has made a small claw mark upon the fascinating world of

---

[106] Bakker, R. T., Sullivan, R. M., Porter, V., Larson, P. and Saulsbury, S.J. (2006). "Dracorex hogwartsia, n. gen., n. sp., a spiked, flat-headed pachycephalosaurid dinosaur from the Upper Cretaceous Hell Creek Formation of South Dakota." in Lucas, S. G. and Sullivan, R. M., eds., Late Cretaceous vertebrates from the Western Interior. New Mexico Museum of Natural History and Science Bulletin 35, pp. 331–345.

dinosaurs. I happen to known more on the subject of paleontology than many might credit, because my eldest daughter was Utahraptor obsessed and I am now living with a passionate Tyrannosaurus-Rex lover, aged three. My credibility has soared within my science-loving family, and I am very much looking forward to reading Dr. Bakker's paper describing 'my' dinosaur, which I can't help visualizing as a slightly less pyromaniac Hungarian Horntail."

Fictional no more, dragons are gaining physical presence. Once we are willing to equate dragons with dinosaurs, and just accept that dinosaur is a relatively new word to describe these creatures we will find numerous examples that reinforce the coexistence of dragons and men. In Utah, several petroglyphs resemble air or land dragons. A pictograph in San Rafael Swell is of something similar to a Pteranodon or Pterodactyl. One in Natural Bridges National Monument looks similar to a sauropod.

Peru is known for dragon and dinosaur-like creatures in their pottery and other artifacts. For example, the Museum of the Nation displays a dragon-like dinosaur on a piece of pottery attributed to the Moche culture (AD 400–1100).

HERE BE DRAGONS 115

Figure 29: Utah Petroglyph

Bishop Bell, who died in 1496, is buried in the foundation of the famous Carlisle Cathedral. The ornate brass engravings around his grave show several animals, some of which appear to be dinosaurs, like a long-neck sauropod and a horned ceratopsian.

Figure 30: Sauropod dinosaurs etched in the brass inlay - Carlisle Cathedral

Chinese dragons, well-known throughout the world, even appear on China's twelve-year calendar cycle. Eleven of these animals are common today (dog, rat, monkey, etc.), so why assume that the twelfth (a dragon) was mythological? The Travels of Marco Polo describes some of these long and lanky "serpents," which included short legs and claws. He claimed the Chinese would use special methods to kill these dragons. Some of the dragons' body parts were used for medicinal purposes, and others were eaten as a delicacy.

In the far north of Queensland, Australia, Aborigines from the Kuku Yalanji tribe described and painted a sea and lake monster that looked surprisingly like a plesiosaur.

Figure 31: Carving Depicting A Dinosaur (Dragon) At Angkor, Cambodia

It seems very clear that dragons, by whatever name they are known, are a genuine part of our past. They appear in our cultural memory through stories and legends, and in scripture. They are part of the creation story. They are part of the flood, and their fossils record their experience during the flood. And they were part of our post-flood experience. Like many other animals, their extinction in no way represents the fact that they did not exist. They are a world-wide memory of the wonder of God's creation.

Figure 32: Chinese Dragon Coin (1906)

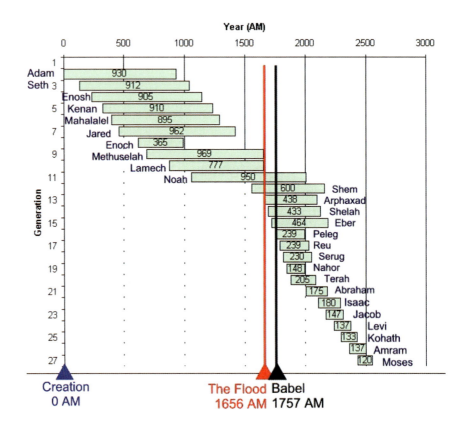

RAIN OF HEAVEN

God's great blessing to the Children of Israel upon leaving Egypt was that the promised land would be "a land of hills and valleys, and drinketh water of the rain of heaven"[107].

In Egypt, for most of the people, the water comes through the Nile river. It rises from below, from beneath their feet. In contrast, in the land of Israel, all precious water comes from above. In Egypt, the people learned to look down for water and all it brought. The natural abundance of water in Egypt along the Nile encouraged the Egyptians to worship the products of their own making. In Israel, and throughout the time spent in the wilderness, the people were taught to look up. First for

---

[107] Deut. 11:11

manna, and forever for water. They were reminded to pray continually, and to remember God's involvement in their lives. In Israel, the people learned to look to God. When they entered the land, He said:

"For the land, whither thou goest in to possess it, is not as the land of Egypt, from whence ye came out, where thou sowedst thy seed, and *wateredst it with thy foot*, as a garden of herbs: But the land, whither ye go to possess it, is a land of hills and valleys, and drinketh *water of the rain of heaven*: A land which the LORD thy God careth for: the eyes of the LORD thy God are always upon it, from the beginning of the year even unto the end of the year."[108]

Throughout most of human history, water has been a metaphor for life. Rain is most frequently associated with the Holy Ghost, revelation, blessings, and God's grace. It is even related to the symbolism associated with direction (south being a symbol for the covenant, or facing God). In fact, in greek, the God of the south wind is Notos (νότος) whose name means "to bring rain."

It is interesting therefore, that this symbol of life, would in the case of Noah, become a symbol for the death and destruction of most of the human race. This seems a conflict, or anachronism,

---

[108] Deut. 11: 10-12, *emphasis added*

but is it? In my experience if there is an apparent contradiction in the behavior of God, I need to reevaluate my assumptions. What assumptions have we made in this analysis? One, that rain is a symbol of life. Two, in the case of Noah, that rain symbolized death. Where is the error?

It may be helpful to look at some ancient Hebrew. This is a pastime of mine, when I am looking for insight into ancient scripture that appears to hold some kind of riddle, or a hidden meaning, I refer to the ancient Hebrew. This can provide additional clues. Hebrew is an amazing language, one that can seem foreign to us not just because of words we do not know, and characters we cannot read, but because it contains layers and layers of meaning. The Jewish Talmud says that "The day when rain falls is as great as the day on which heaven and earth were created."[109] It is clear from this reference, that an understanding of rain was important to the Jews.

Remember what Joseph Smith said about his use of the English language (he was also a lover of Hebrew):

> "Oh Lord God, deliver us from this prison, almost as it were, of paper, pen and ink, and of a crooked, broken, scattered and imperfect language."

---

[109] Masehet Ta'anit, 8b

Joseph's broken, imperfect language was English. He wrote this in letter to William W. Phelps in 1832, specifically referring to the frustrations he felt in using the English language to communicate subtle and specific ideas.

Many references to the blessings of "rain" are actually associated with "tal", most easily translated as dew, the droplets of water that form through condensation upon the ground or plants. In dry seasons, this dew can drench the soil and plants. It's value is expressed in this beautiful statement in the Talmud: "A city that enjoys more dew than other cities, enjoys more fruits." While Jewish prayers for rain are only offered in the winter, daily prayers for "tal", or dew, are offered throughout the year.

Like many words used in our language, what we call simply "rain" has several additional words in Hebrew, each with a meaning richer and more descriptive than our simple "rain". For the winter rain, these include "yoreh"; early (or autumn) rain, "malkosh"; late (or spring) rain, and plain old "matar"; from which we get the modern Hebrew word for umbrella, "mitriyah", and "geshem", or a strong rain.

It is interesting then, that when geshem appears in Scripture, it is usually associated with a heavy rain. The word geshem materializes twice in the story of Noah's flood. In addition, we have geshem nedavot, "bounteous rain," in Psalms, geshem gadol, "a big rain," and hamon ha-geshem, "abundant rain," in the Book of Kings, and geshem shotef, "torrential rain," three times in the Book of Ezekiel.

A Jewish rabbi describes the rains in the days of Noah this way: "...when He brought them [the rains] down, He brought them down with mercy, so that if they [the people] would repent, they [the rains] would be rains of blessing. When they did not repent, they became a flood."[110]

Jewish culture says that the scarcity of rain in Israel is a spiritual safeguard. Asking for rain, praying for it, reinforces the bond with God continually. As they entered the promised land, the Israelites were warned about the danger of thinking themselves too responsible for their own successes:

> "Beware that thou forget not the LORD thy God, in not keeping his commandments, and his judgments, and his statutes, which I command thee this day: Lest when thou hast eaten and art full, and hast built goodly houses, and dwelt therein; And when thy herds and thy flocks multiply, and thy silver and thy gold is multiplied, and all that thou hast is multiplied; Then thine heart be lifted up, and thou forget the LORD thy God, which brought thee forth out of the land of Egypt, from the

---

[110] Rabbi Shlomo Yitzchaki (Rashi), *The Torah with Rashi's Commentary*, Mesorah Publications, 1999

house of bondage; Who led thee through that great and terrible wilderness, wherein were fiery serpents, and scorpions, and drought, *where there was no water*; who brought thee forth *water out of the rock of flint*; Who *fed thee in the wilderness with manna*, which thy fathers knew not, *that he might humble thee*, and that he might prove thee, to do thee good at thy latter end; And thou say in thine heart, My power and the might of mine hand hath gotten me this wealth."[111]

Failure to remember the Lord, and the blessings he provides, including life-giving rain, has a consequence, which is revealed in the versus that follow:

"And it shall be, if thou do at all forget the LORD thy God, and walk after other gods, and serve them, and worship them, I testify against you this day that ye shall surely perish. As the nations which the LORD destroyeth before your face, so shall ye perish; because ye would not be obedient unto the voice of the LORD your God."[112]

There is a wonderful Jewish tale of Honi the Circle Drawer told in the Talmud. Honi Ha-me'aggel (the Circle-Maker) was a renowned pietist in the period of the Second Temple (first century BCE) who was said to have performed good deeds by

---
[111] Deut. 8:11-17, *italics added*
[112] Deut. 8:19-20

using extraordinary powers of prayer or by performing miracles. According to popular legend, Honi slept for seventy years and on awakening prayed for death rather than living in a strange world.

The following story, which tells of his power to bring rain in times of drought, is recorded in many sources. His name, ha-Me'aggel ("one who draws circles") is usually connected with this incident. Some scholars claim he was named Honi Ha-me'aggel after the place from which he came, while others suggest he was so called as he was often called to repair roofs or ovens, with a ma'gillah ("roller"):

"Once there was a terrible drought in the land of Israel. It was already the month of Adar, which usually marks the end of the rainy season and the beginning of spring, but no rains had fallen all winter long.

"So the people sent for Honi the Circle-Maker. He prayed, but still no rains came. Then he drew a circle in the dust and stood in the middle of it. Raising his hands to heaven, he vowed, "God, I will not move from this circle until You send rain!"

"Immediately a few drops fell, hissing as they struck the hot white stones. But the people complained to Honi, "This is but a poor excuse for rain, only enough to release you from your vow."

"So Honi turned back to heaven and cried, "Not for this trifling drizzle did I ask, but for enough rain to fill wells, cisterns, and ditches!

Then the heavens opened up and poured down rain in buckets, each drop big enough to fill a soup ladle. The wells and the cisterns overflowed, and the wadis flooded the desert. The people of Jerusalem ran for safety to the Temple Mount.

""Honi!" they cried. "Save us! Or we will all be destroyed like the generation of the Flood! Stop the rains!"

"Honi said to them, "I was glad to ask God to end your misery, but how can I ask for an end to your blessing?"

"The people pleaded with him, and he finally agreed to pray for the rain to stop. "Bring me an offering of thanksgiving," he told them, and they did.

"Then Honi said to God, "This people that You brought out of Egypt can take neither too much evil nor too much good. Please give them what they ask so that they may be happy."

"So God sent a strong wind that blew away the fierce rains, and the people gathered mushrooms and truffles on the Temple Mount.

"Then Shimon ben Shetakh, head of the Sanhedrin in Jerusalem, said to Honi, "I should excommunicate you for your audacity, but how can I, since you're Honi! God coddles you as a father does his young child. The child says: 'Hold me, Daddy, and bathe me, and give me poppyseeds and peaches and pomegranates,' and his father gives him whatever he wants."

"So it was with Honi the Circle-Maker."[113]

How does this help to resolve our dilemma? Rain is typically prayed for, and considered a blessing. With this background, let's reconsider the rain. In particular, remember the words of Nephi:

"He doeth not anything save it be for the benefit of the world; for he loveth the world, even that he layeth down his own life that he may draw all men unto him. Wherefore, he commandeth none that they shall not partake of his salvation."[114]

With this point of reference, we understand that whatever God does, he does for our benefit, because he loves us. How was the rain, sent in such quantity that it silenced the lives of all those living, only sparing Noah's family, for our benefit?

President John Taylor said that "by taking away their earthly existence [God] prevented them from entailing their sins upon

---

[113] *The Classic Tales: 4,000 Years of Jewish Lore*, Ed. Ellen Frankel, NJ: Jason Aronson Inc., 1989
[114] 2 Nephi 26:24

their posterity and degenerating them, and also prevented them from committing further acts of wickedness"[115].

Those who were being born into such wickedness would not have the opportunity to choose. Darkness dominated, with only a flicker of light being kept alive by Noah's family. And for Noah, even that was being threatened, as Noah's granddaughters were falling under the influence of the world.

A loving Heavenly Father could not continue to send children into these wicked circumstances. As Elder Neal A. Maxwell said, "when corruption had reached an agency-destroying point that spirits could not, in justice, be sent here"[116]. Even Noah had recognized that there was no safety for any of God's children when he "repented that the Lord had made them."

But it also benefited the wicked, preventing them from continuing in their path. "For Christ also hath once suffered for sins, the just for the unjust, that he might bring us to God, being put to death in the flesh, but quickened by the Spirit: By which also he went and *preached unto the spirits in prison*; Which sometime were disobedient, when once the longsuffering of God waited *in the days of Noah*, while the ark

---

[115] Discourse Delivered by President John Taylor, *Deseret News*, Jan. 16, 1878, 787
[116] We Will Prove Them Herewith, Neal A. Maxwell, 1982, 58

was a preparing, wherein few, that is, eight souls were saved by water."[117]

When Christ visited those in spirit prison, after he was crucified, the power of his atonement reached out very pointedly to those who were disobedient in the days of Noah.

So, there is no conflict. Rain is a blessing, something to be prayed for. But, we may not recognize blessings when we see them. We may be so entrenched in thinking something is unfair, or too painful, when in reality, it is the work of a loving Heavenly Father, always considering our good, "for he loveth the world."

---

[117] 1 Peter 3:18-20

130 THE TOKEN OF THE BOW

"Inside Out Rainbow", Craig Harms

A FLOOD, A CITY, AND A GARDEN

We remember that Enoch was the patriarch of a great city, Zion. Out of respect for him, we often call it the City of Enoch. A city that was so righteous, it was taken up to Heaven. But who was this man? We must be grateful to the Prophet Joseph for the restoration of so many lost truths related to Enoch. He's a lot more like someone you could relate to, I think.

The Lord called Enoch out of a wicked world, but Enoch was afraid. He said to the Lord that he was "but a lad, and all the people hate me; for I am slow of speech..."[118]

The Lord told Enoch that he would protect him, that no one would have the ability to take his life, "no man shall pierce

---

[118] Moses 6:31

thee", he told him. He also removed his speech impediment, saying to him: "Open thy mouth, and it shall be filled, and I will give thee utterance."[119]

The Lord removed these obstacles from him, and here we see the pattern of Enoch, because the Lord protected him, and filled his mouth with what he needed to say. He lifted the burden.

It is written that Enoch "walked with God," as did his people, and he talked with God, "even as a man talketh one with another, face to face"[120]. Enoch had work to do.

As you look around, you can't help but see the wickedness in the world you live in. Enoch saw this, and much more. The Lord gave him visions. He saw all the children of men, and "all their doings" and Moses 7:44 says that Enoch wept and in one of the most touching scriptures, refused to be comforted.

The Lord told him not to weep, but to look up. When he looked up, what was it that brought comfort to Enoch? "Enoch saw the day of the coming of the Son of Man, even in the flesh; and his soul rejoiced." Seeing the Lord redeem his people brought

---

[119] Moses 6:32
[120] Genesis 5:24; Moses 7:4; Doctrine & Covenants 84:19–24; Doctrine & Covenants 107:49; Moses 6:39; Moses 7:20, 69

Enoch the only comfort that would satisfy. Once again the Lord removed the burden, through His atonement.

The City of Enoch was not a place Enoch went to save a people. It was a place a saved people built. They gathered together because of their belief, and their love. This belief and love made a Zion. Rather than building a place, Enoch and the Lord built a people.

Enoch then did as the Lord asked him, and preached to his people. Many received his words, and became a righteous people. They were so righteous, that as the world became more and more wicked, and as their times became more and more troubled, the Lord took Zion to himself. They were removed from their troubles, through the power of the Lord.

It is interesting that Enoch's son did not go with him and the city of Zion. He was certainly a righteous man. Why wouldn't he go? His name was Methuselah, and scripture tells us:

"And it came to pass that Methuselah, the son of Enoch, was not taken, that the covenants of the Lord might be fulfilled, which he made to Enoch; for he truly covenanted with Enoch that Noah should be of the fruit of his loins."[121]

---

[121] Moses 8:2

Enoch was removed, but Methuselah was given the task to endure.

Things didn't get better with Zion removed. In fact, they continued to get worse. So much so, that scripture tells us that someone repented of creating the world. Genesis would seem to say that the Lord repented of creating man, but thankfully Joseph Smith corrected this misconception, and lets us know that it was Noah that was so disappointed that he wished God had not created man. Why?

As we have discussed earlier, Noah's sons, Japeth, Ham and Shem, were righteous. But the Lord says that Noah's granddaughters "have sold themselves." The wickedness was entering into the lives of Noah's family. And he was afraid. He tried to teach the people, but they would not listen. So God sent a flood to destroy wickedness from the world, to save the children that would be brought into it. But he told Noah to build an ark.

This is the pattern of Noah. Sometimes the Lord simply gives us the ability to endure the trial. He will give us strength, knowledge, or endurance.

Several years ago, I was given the opportunity, with my sweet wife, to participate in an activity with our Stake. It was a reenactment of the hand cart trek of the pioneers across the plains. We were fortunate that our group would actually walk in the steps of these early pioneers, by following their path through Wyoming. We were excited to be asked to serve as support staff for the trip.

My wife, Carol, is a walker. She loves to walk, and decided we should prepare for the trip by building our stamina, and doing some hiking. As we began this process, I found something out about myself. I was no longer the young man who could run for miles. In fact, after just a mile or two of walking, my back began to ache. As we continued, it hurt even more. I was grateful that we were just support staff, and I would likely be able to ride in the support vehicles with the other Stake leaders.

Just a few weeks before our expected departure, we received a call from the Stake leaders. They told us that another couple would not be able to go as a "ma and pa" for their group (a back problem from the pa I expect), and a replacement was needed. The question came: "Would we be willing to take their place?"

We answered, with faith, that we would be willing. I was still concerned, but thought, "How hard could it be?"

As the day came for our trek, the excitement overwhelmed any concerns I had. Upon arriving at the site, we loaded up our hand cart and began the nine mile hike to our camp site. My initial exuberance was shortly overcome by twinges of pain in my back.

At two miles into the hike the pain was distracting. By four miles it was irritating, and by five miles overwhelming. But, surprisingly, I was able to complete the hike. But...we still had tents to set up. And so we did.

By the time the last tent was assembled, I crawled into our tent and lay on the ground trying to will the pain away. It would not relent.

Then a voice came at the tent door. "Would we be willing to help setup the tents for the group behind us? They were delayed on the trail, and rain was coming. They didn't want their things to get wet, so could we assemble their tents and move their things inside?"

We gathered the youth, and began the work. By the time I limped back to our camp, I could only lay on the ground in our tent, and try not to cry out.

My good wife asked if someone could help me. Shortly Gordon Smith, a member of the Stake Presidency, came to our tent and asked if I wanted a blessing. I readily agreed.

As I sat there, with his hands on my head, I waited for words of healing. I wanted to hear the words "take up thy bed and walk."

What I heard instead brought a sinking feeling. He pronounced a blessing on me and told me "you will have the ability to endure." I didn't want to endure, I wanted to be healed. But endurance was what the Lord offered me. I was given the blessing of Noah, the ability to bear the burden. I wanted to be Enoch and have the trial taken from me.

Each day for the next week, we hiked from two to five miles, and each day I would hunch over in pain through the hike.

Finally the day came to hike back to our buses. The youth loaded things on the carts, and were preparing to begin the nine miles out. A member of the Stake came to me and asked if I would like to ride out with one of them. They knew it had been a painful week, and maybe getting a ride back would help. I considered this offer, and then remembered the blessing I had been given. I was blessed to endure, not to escape. I thanked them for the offer, and began the hike with my adopted family. Once again, the pain began to build. Step by step it increased. Mile by mile, my discomfort grew. Finally, just as I thought I could bear no more, one of the youth yelled out that the cart was stuck in the sand. They were trying and trying to push it out. I couldn't just watch them struggle, so I leaned, put my hands on the back of the cart, and pushed.

The pain left. I discovered that as long as I leaned into the cart and pushed, I had absolutely no pain. All I had to do was fully commit. Needless to say, I pushed that cart all the way back to the bus. The Lord blessed me, and let me be an Enoch, if just for a moment.

Just as with Noah and Enoch, my rescue was a personal event. Yours will be too. It likely will affect many (your families, and generations to come), but it occurs in a personal and individual way. Though salvation is available to everyone, our relationship with the Savior is an intimate and personal one. We must each choose to surrender our will to trust in our Savior Jesus Christ.

Elder Whitney (as a young missionary), had the following powerful dream:

> "One night I dreamed … that I was in the Garden of Gethsemane, a witness of the Savior's agony. … I stood behind a tree in the foreground. … Jesus, with Peter, James, and John, came through a little wicket gate at my right. Leaving the three Apostles there, after telling them to kneel and pray, He passed over to the other side, where He also knelt and prayed … : 'Oh my Father, if it be possible, let this cup pass from me; nevertheless not as I will but as Thou wilt.'
>
> "As He prayed the tears streamed down His face, which was [turned] toward me. I was so moved at the sight that I wept also, out of pure sympathy with His great sorrow. My whole heart went out to Him. I loved Him with all

my soul and longed to be with Him as I longed for nothing else.

"Presently He arose and walked to where those Apostles were kneeling—fast asleep! He shook them gently, awoke them, and in a tone of tender reproach, untinctured by the least show of anger or scolding, asked them if they could not watch with Him one hour. ...

"Returning to His place, He prayed again and then went back and found them again sleeping. Again He awoke them, admonished them, and returned and prayed as before. Three times this happened, until I was perfectly familiar with His appearance—face, form, and movements. He was of noble stature and of majestic mien ... the very God that He was and is, yet as meek and lowly as a little child.

"All at once the circumstance seemed to change. ... Instead of before, it was after the Crucifixion, and the Savior, with those three Apostles, now stood together in a group at my left. They were about to depart and ascend into heaven. I could endure it no longer. I ran from behind the tree, fell at His feet, clasped Him around the knees, and begged Him to take me with Him.

"I shall never forget the kind and gentle manner in which He stooped and raised me up and embraced me.

It was so vivid, so real that I felt the very warmth of His bosom against which I rested. Then He said: 'No, my son; these have finished their work, and they may go with me; but you must stay and finish yours.' Still I clung to Him. Gazing up into His face—for He was taller than I—I besought Him most earnestly: 'Well, promise me that I will come to You at the last.' He smiled sweetly and tenderly and replied: 'That will depend entirely upon yourself.' I awoke with a sob in my throat, and it was morning."

"Because of the eternal sanctity of man's agency upon which this mortal life was founded, the Savior cannot take from us our will. But, once we enter into the sacred covenant of baptism, He is able to pull us back to him, if we just ask. The Savior stands beside us waiting to heal our wounds and to lift us into eternal salvation, to rescue us, but He can only do that with our invitation. We must choose Him. For us, there is only one plan of rescue; it is in and through His atoning sacrifice. He descended below all things to rescue us.

"What do you think? If a man has a hundred sheep, and one of them has gone astray, does he not leave the ninety-nine on the mountains and go in search of the one that went astray? And if he finds it, truly, I say to you, he rejoices over it more than over the ninety-nine that never went astray"[122].

"Though it involves sorrow, and the pain of being lost, repentance is ultimately about finding joy, and when we return to God, we have reason to rejoice. The Good Shepherd says, "Rejoice with me, for I have found my sheep that was lost. Just so, I tell you, there will be more joy in heaven over one sinner who repents than over ninety-nine righteous persons who need no repentance"[123].

The Son of Man came to seek and to save the lost:

"For thus says the Lord GOD: Behold, I, I myself will search for my sheep and will seek them out... I myself will be the Shepherd of my sheep, and I myself will make them lie down, declares the Lord GOD. I will seek the lost, and I will bring back the banished, and I will bind up the injured, and I will strengthen the sick"[124].

President Dieter F. Uchtdorf said, "We acknowledge that your path will at times be difficult. But I give you this promise in the name of the Lord: rise up and follow in

---

[122] Matthew 18:12-13
[123] Luke 15:6-7
[124] Ezekiel 34:11,5-16

the footsteps of our Redeemer and Savior, and one day you will look back and be filled with eternal gratitude that you chose to trust the Atonement and its power to lift you up and give you strength."

And finally, from the Prophet, Thomas S. Monson:

"We were not placed on this earth to walk alone.

"What an amazing source of power, of strength, and of comfort is available to each of us. He who knows us better than we know ourselves, He who sees the larger picture and who knows the end from the beginning, has assured us that He will be there for us to provide help if we but ask.

"We have the promise: 'Pray always, and be believing, and all things shall work together for your good.'"[125]

---

[125] Doctrine & Covenants 90:24

THE AMAZING REACH OF COVENANTS

Nothing has amazed and comforted me more than the realization that I have underestimated the breadth, length and reach of the Lord's covenants.

While much of our focus in this book will be on the covenant the Lord made with Noah, we'll first look at a few others. Beginning, well, at the beginning.

When Adam learned that he would be leaving the Garden, I can only imagine the loss he felt. We have no idea how long he and Eve lived in the Garden, but it was likely some time. It's easy to continue to follow the Lord's "day" references, and consider that things moved along pretty quickly. I, however, believe they spent quite some time there. They named all the animals, something that likely took some time, and can be associated with developing their language. There was sufficient time that the serpent's temptations had time to work their way into their minds. The uncanonized *Book of Jubilees* says it was seven years, but it was certainly more than six days (the period of creation) and less than 130 years (Adam's age when Seth was born[126]). At any rate, the garden had become home. They had frequent visits from the Lord, and were enjoying all the blessings of their new unfallen physical bodies. With one decision, one act, all this was changing. They were leaving the garden, in fact they were being "driven" from it.[127] I can feel their loss. It must have been nearly unbearable for them.

I am certain they wanted nothing more than to return. I can imagine their looks back, longingly, toward the garden and all it represented: home, comfort, a closeness to the Lord, peace. It was all there, and it was all behind them…until the Lord made a covenant with them.

There's an amazing consistency with covenants. Over and over we see the deepest desires of our hearts being fulfilled in the promises of the covenants we make with the Lord. In this case, the Lord promised Adam and Eve that they would have what they desired: to return to His presence, to have everything they

---

[126] Genesis 5:3
[127] Moses 4:31, Genesis 3:24

had lost restored[128]. Further, through the covenant of baptism, they would become Christ's children, and would then be heirs to all He possessed, most significantly, His Father's likeness[129].

In itself this would be a wonderful fact, but thankfully, there's more to the story. Because, the blessings of this covenant were not limited to Adam but extended to all his descendants. Think of it, Adam and all his children, and their children, and on and on, each becoming heirs to the promise. This covenant extends to all, to us.

This is the lost treasure of the Lord's covenants, that they reach out to each of us. The Lord's pattern is to bless those he loves by blessing all their posterity.

It doesn't apply just to Adam. Consider Abraham. He wanted so much to have a family, but he had not been able to have any children. He was already an old man when he finally had a son with his beloved Sarah. All his life he had been a stranger in a strange land, a nomad. He wanted a home. And when the Lord covenanted with Abraham, what was the nature of this covenant? He would have a posterity more numerous than the sands of the sea, and would have a land of his own, an amazing land of promise.

---

[128] Genesis 1:28-30
[129] Moses 6:65-68

But, once again, this covenant was to be with Abraham and his descendants to the end of time. This covenant was renewed with Isaac, and again with Jacob. We are each heirs to it. We will each have numberless posterity, and a promised land of our own. The blessing of Abraham reaches out to each of us.

But, this is a book about Noah. What were the blessings promised to Noah? The Lord covenanted with Noah that he would never again destroy all living things by a flood, and renewed the covenant that the earth is ours. There is an interesting mention of the fact that the animals will fear man[130]. The earlier covenant made to Adam states that the animals will be subject to man. Noah is now told to eat of the flesh of the animals[131]. Most importantly, he is promised that through his faithfulness, the Earth would be renewed.

How does all of this affect us? As we consider the reach of these great covenants, we begin to recognize a pattern.

When the Lord covenants with these patriarchs, the promises have threads that reach to all those who follow. Why would this pattern be broken in dealing with us? It isn't! When you enter into covenants, listen closely, the echoes of this pattern are there. Not only are you to be blessed for your obedience, but those who follow after you. The Lord wants to bless his

---
[130] Genesis 9:2
[131] Genesis 9:3

children! So much so, that the righteousness of his servants reaches out to those around them. This is the miracle of the covenant.

When the Lord tells his servants that they are to be a light that is shed in the darkness[132], it is not just to illuminate them, but those who follow! When he tells his servants that they are to be the leaven in the bread[133], it is to lift the entire loaf! When he compares his servants to salt[134], it is not just to preserve themselves, but everyone!

The covenants made to Adam, Abraham, Moses, Noah and Joseph extend to you. The covenants you make extend to your families, and those around you.

We often consider covenants in light of our employment relationships. When we do a good job, or meet our end of the employment contract, we receive the agreed-upon payment. It makes a sort of sense. In a covenant, we promise to do certain things, and in exchange God rewards, or pays us. But in a job, all that is at stake if we don't fulfill our obligation is a reduction in our payment. We can always work some overtime and make it up, or get a new job if we tire of the old one. You can see how

---

[132] John 1:5
[133] Matthew 13:33
[134] Matthew 5:13

this can be problematic if we apply it to our relationship to the Lord.

I prefer to look at a covenant as a map. When we want to arrive at a specific destination, it is helpful to have a set of instructions. Following in the steps of someone who has travelled the route before (the map maker), gives us confidence that the destination is sure. When we analyze a covenant, whether one we have personally entered into, or one we inherit, we can see the elements of this road map. It is more a cause and effect relationship. So, what is the destination of this covenant road map?

I often hear people say that the goal of this life is to return to our Father in Heaven. I believe this is a part of why we are here. But it misses the fact that before we came to this earth we were already with him. We already lived in his presence, and yet we were willing, even anxious, to leave for a season. So, there must be something more.

Heavenly Father says that His goal is our immortality and eternal life[135]. So, the oft stated goal (to return to Heavenly Father) achieves half of the Lord's goal for us. Because of the Atonement, immortality is already achieved. But eternal life,

---

[135] Moses 1:39

that's another matter. It has been said that immortality is about quantity, while eternal life is about quality.

So, when we left Heavenly Father's presence, it wasn't simply about coming back. It was about coming back with a higher quality of life, a life like his. We didn't simply want to come back, we wanted to learn how to be like him. We wanted a body, like his, but first we needed to have one like ours.

Now here's where we tie all this talk about covenants, and threads, and immortality, and road maps, and eternal life together. When we enter into a covenant, it is receiving the directions, the road map, that will allow us to achieve what Heavenly Father promised us: the desire of our hearts, to be like him. And, because of the nature of covenants, we extend that promise to those we love, which ultimately is everyone.

This is the amazing reach of covenants...that not only do we become more like Him, but our families do as well. That is the power of the plan. The Lord is looking for every opportunity to help us achieve what we always wanted, the desire of our hearts: to have all that he has. This is not just a story of quantity, but of quality.

"Jamey's Rainbow"

Jamey Scott (1984-1994), son of Sue and Craig Harms

WHO SAYS THERE WAS A FLOOD ANYWAY?

A library of books have been written arguing both sides of this argument. But, here we turn our attention to the words of those who speak with divine authority, or insight. While science may be confused, and paradigms may change, these stay true and consistent.

In other chapters, a discussion of the world-wide nature of the records pertaining to the flood is reviewed in detail, but an example is offered here. The Flood account is mirrored in an ancient Chinese record that states "the earth fell to pieces and water rushed up from its bosom, overflowing the earth. The Sun, Moon and stars changed their place in the heavens". The

Chinese, as well as the Koreans, trace their lineage back to Noah through Lo-Shen, or Shem.

No historical event is so well recorded throughout the world. While many scholars attempt to put the cart before the horse, claiming that many of these myths or legends that exist in other cultures serve as the basis for the scriptural narrative, the reality is something much more straightforward: the flood actually happened, the story was passed down to all of Noah's descendants, but many did not carry with them the written records of Noah (which included Adam's record as well as Enoch's). As a result of losing the record, these groups were forced to rely on oral traditions, which become corrupted through years of retelling. What we should be doing is looking for the threads of truth that connect them all.

Years ago I served aboard a US Navy ship. Each division on the ship took turns sending a senior petty officer to serve in the ships police force (the Master At Arms). When it was my division's turn, they chose to send me. I would serve six months as a ship's security officer.

One of my first assignments was to investigate an altercation which occurred ashore, involving a number of the ship's personnel. It required me to interview over sixty individuals. After completing these interviews, I was amazed to find that I had sixty unique narratives, each ostensibly describing the same event. It came as a surprise to me. So much so, that I spoke with the Master Chief who was in charge. He shared something with me that has served me well in the thirty-plus years since. He taught me how to first find the things I knew to be true, that were independent of any witness Then to determine what consistent elements existed in the majority of statements I had taken. And then, finally, to decide which other accounts or events most logically connected the know facts and the common remembrances. These events would constitute the most likely version of the facts.

The stories, legends and remembrances of a global flood can be put to the same test. What is revealed is compelling. If we assume that each of these stories is a unique version of a common event, we can determine what actually happened, and then compare it to the account given by Moses (a trustworthy first-hand witness, even several thousand years after the event).

To simplify the process, I will pick some representative stories. But, understand, that every culture I have researched contains a flood narrative. They all learned from the ancestors that such an event occurred.

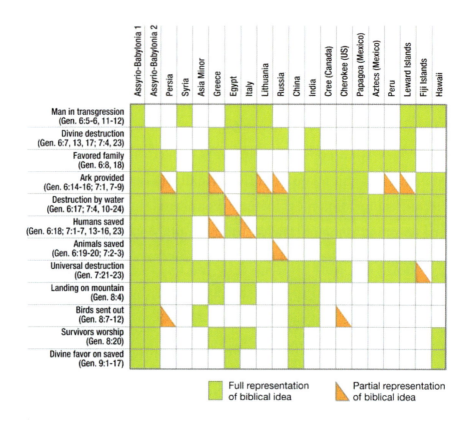

Africa: Southwest Tanzania

Once upon a time the rivers began to flood. The god told two people to get into a ship. He told them to take lots of seed and to take lots of animals. The water of the flood eventually covered the mountains. Finally the flood stopped. Then one of the men, wanting to know if the water had dried up let a dove loose. The dove returned. Later he let loose a hawk which did not return. Then the men left the boat and took the animals and the seeds with them.

Asia: China

The Chinese classic called the Hihking tells about "the family of Fuhi," that was saved from a great flood. This ancient story tells that the entire land was flooded; the mountains and everything, however one family survived in a boat. The Chinese consider this man the father of their civilization. This record indicates that Fuhi, his wife, three sons, and three daughters were the only people that escaped the great flood. It is claimed, that he and his family were the only people alive on earth, and repopulated the world.

Figure 33: The Chinese Character for Boat

Babylon

Gilgamesh met an old man named Utnapishtim, who told him the following story. The gods came to Utnapishtim to warn him about a terrible flood that was coming. They instructed

Utnapishtim to destroy his house and build a large ship. The ship was to be 10 dozen cubits high, wide and long. Utnapishtim was to cover the ship with pitch. He was supposed to take male and female animals of all kinds, his wife and family, provisions, etc. into the ship. Once ship was completed the rain began falling intensely. The rain fell for six days and nights. Finally things calmed and the ship settled on the top of Mount Nisir. After the ship had rested for seven days Utnapishtim let loose a dove. Since the land had not dried the dove returned. Next he sent a swallow which also returned. Later he let loose a raven which never returned since the ground had dried. Utnapishtim then left the ship.

Chaldean

There was a man by the name of Xisuthrus. The god Chronos warned Xisuthrus of a coming flood and told him to build a boat. The boat was to be 5 stadia by 2 stadia. In this boat Xisuthrus was to put his family, friends and two of each animal (male and female). The flood came. When the waters started to recede he let some birds loose. They came back and he noticed they had mud on their feet. He tried again with the same results. When he tried the third time the birds did not return. Assuming the water had dried up the people got out of the boat and offered sacrifices to the gods.

India

A long time ago lived a man named Manu. Manu, while washing himself, saved a small fish from the jaws of a large fish. The fish told Manu, "If you care for me until I am full grown I will save you from terrible things to come". Manu asked what kind of terrible things. The fish told Manu that a great flood would soon come and destroy everything on the earth. The fish told Manu to put him in a clay jar for protection. The fish grew and each time he outgrew the clay jar Manu gave him a larger one. Finally the fish became a ghasha, one of the largest fish in the world. The fish instructed Manu to build a large ship since the flood was going to happen very soon. As the rains started Manu tied a rope from the ship to the ghasha. The fish guided the ship as the waters rose. The whole earth was covered by water. When the waters began subsiding the ghasha led Manu's ship to a mountaintop.

Australia

There is a legend of a flood called the Dreamtime flood. Riding on this flood was the woramba, or the Ark Gumana. In this ark was Noah, Aborigines, and various animals. This ark eventually came to rest in the plain of Djilinbadu where it can still be found. They claim that the white mans story about the ark landing in the middle east is a lie that was started to keep the aborigines in subservience. This legend is undoubtedly the product of aboriginal legends merging with those of visiting missionaries, and there does not appear to be any native flood stories from Australia.

Europe: Greece

A long time ago, perhaps before the golden age was over, humans became proud. This bothered Zeus as they kept getting worse. Finally Zeus decided that he would destroy all humans. Before he did this; Prometheus, the creator of humans, warned his human son Deucalion and his wife Pyrrha. Prometheus then placed this couple in a large wooden chest. The rains started and lasted nine days and nights until the whole world was flooded. The only thing that was not flooded was the peaks of Mount Parnassus and Mount Olympus. Mount Olympus is the home of the gods. The wooden chest came to rest on Mount Parnassus. Deucalion and his wife Pyrrha got out and saw that everything was flooded. They lived on provisions from the chest until the waters subsided. At Zeus' instruction they re-populated the earth.

North America: Mexico

The Toltec natives have a legend telling that the original creation lasted for 1716 years, and was destroyed by a flood and only one family survived.

Aztec- A man named Tapi lived a long time ago. Tapi was a very pious man. The creator told Tapi to build a boat that he would live in. He was told that he should take his wife, a pair of every animal that was alive into this boat. Naturally everyone thought he was crazy. Then the rain started and the flood came. The men and animals tried to climb the mountains but the mountains became flooded as well. Finally the rain ended. Tapi decided that the water had dried up when he let a dove loose that did not return.

United States

The Ojibwe natives who have lived in Minnesota since approximately 1400AD also have a creation and flood story that closely parallels the Biblical account. "There came a time when the harmonious way of life did not continue. Men and women disrespected each other, families quarreled and soon villages began arguing back and forth. This saddened Gitchie Manido [the Creator] greatly, but he waited. Finally, when it seemed there was no hope left, Creator decided to purify Mother Earth through the use of water. The water came, flooding the Earth, catching all of creation off guard. All but a few of each living thing survived." Then it tells how Waynaboozhoo survived by floating on a log in the water with various animals.

Ojibwe - Ancient native American creation story tells of world wide flood.

Delaware Indians - In the pristine age, the world lived at peace; but an evil spirit came and caused a great flood. The earth was submerged. A few persons had taken refuge on the back of a turtle, so old that his shell had collected moss. A loon flew over their heads and was entreated to dive beneath the water and bring up land. It found only a bottomless sea. Then the bird flew far away, came back with a small portion of earth in its bill, and guided the tortoise to a place where there was a spot of dry land.

South America: Inca

During the period of time called the Pachachama, people became very evil. They got so busy coming up with and performing evil deeds they neglected the gods. Only those in the high Andes remained uncorrupted. Two brothers who lived in the highlands noticed their llamas acting strangely. They asked the llamas why and were told that the stars had told the llamas that a great flood was coming. This flood would destroy all the life on earth. The brothers took their families and flocks into a cave on the high mountains. It started to rain and continued for four months. As the water rose the mountain grew keeping its top above the water. Eventually the rain stopped and the waters receded. The mountain returned to its original height. The shepherds repopulated the earth. The llamas remembered the flood and that is why they prefer to live in the highland areas.

A summary of these accounts reveals a yes answer to some common flood story questions:

Is there a favored family? 88%

Were they forewarned? 66%

Is flood due to wickedness of man? 66%

Is catastrophe only a flood? 95%

Was flood global? 95%

Is survival due to a boat? 70%

Were animals also saved? 67%

Did animals play any part? 73%

Did survivors land on a mountain? 57%

Was the geography local? 82%

Were birds sent out? 35%

Was the rainbow mentioned? 7%

Did survivors offer a sacrifice? 13%

Were specifically eight persons saved? 9%

But, most compelling of all is the fact that prophets through the ages have identified Noah as a historical, not a mythical, character. These include Enoch[136], Abraham[137], the Apostle Paul[138], Amulek[139], Moroni[140], Matthew[141], Peter[142], Joseph Smith, Jr.[143], and Joseph F. Smith[144]. Jesus Christ himself spoke to the Nephites of the "waters of Noah"[145].

Joseph Smith not only believed the flood of Noah occurred as the Bible says it did, but related it to the second coming of Christ. As is often the case with prophets after the flood, it became a reference point for events of the last days, and of the Savior's return.

Joseph said:

---

[136] Moses 7:42–43
[137] Abraham 1:19
[138] Hebrews 11:7
[139] Alma 10:22
[140] Ether 6:7
[141] Joseph Smith Matthew 1:41–42 and Matthew 24:37-39
[142] II Peter 2:5
[143] Doctrine & Covenants 84:14–15; Doctrine & Covenants 133:54
[144] Doctrine & Covenants 138:9, 41
[145] 3 Nephi 22:9

"I have asked of the Lord concerning His coming; and while asking the Lord, He gave a sign and said, "In the days of Noah I set a bow in the heavens as a sign and token that in any year that the bow should be seen the Lord would not come; but there should be seed time and harvest during that year: but whenever you see the bow withdrawn, it shall be a token that there shall be famine, pestilence, and great distress among the nations, and that the coming of the Messiah is not far distant."[146]

Right before His death, Christ told His disciples details about His return. He said:

"For as were the days of Noah, so will be the coming of the Son of Man. For as in those days before the flood they were eating and drinking, marrying and giving in marriage, until the day when Noah entered the ark, and they were unaware until the flood came and swept them all away, so will be the coming of the Son of Man."[147]

He describes those living in Noah's day as "unaware" that a flood was coming. Christ says the flood swept them all away. The fact that Christ is linking an event where "all flesh died that moved upon the earth, both of fowl, and of cattle, and of beast,

---

[146] Joseph Fielding Smith, *Teachings of the Prophet Joseph Smith*, Section 6, 1843-44, p341
[147] Matthew 24:37-39

and of every creeping thing that creepeth upon the earth, and every man"[148] to His eventual return, means something about that return. It's going to be big, and it's going to affect everyone, *just as the flood did.*

According to Jesus Christ, Noah, the ark, and a global flood that killed all the animals and people on the earth, were as historically real as His second coming.

One of the most compelling modern recitals of the Noah story comes from a modern prophet, Spencer W. Kimball. He wrote:

"Paul speaking to the Hebrews said:

"'By faith Noah, being warned of God of things not seen as yet, moved with fear, prepared an ark to the saving of his house.'[149]

"As yet there was no evidence of rain and flood. His people mocked and called him a fool. His preaching fell on deaf ears. His warnings were considered irrational. There was no precedent; never had it been known that a deluge could cover the earth. How foolish to build an ark on dry ground with the sun shining and life moving forward as usual! But time ran out. The ark was finished. The floods came. The

---

[148] Genesis 7:21
[149] Hebrews 11:7

disobedient and rebellious were drowned. The miracle of the ark followed the faith manifested in its building."[150]

The flood was obviously a very real event to this Prophet of the Lord.

For us to deny the authenticity of this event, is to deny the validity of the words of those we trust on so many other subjects. How can we discard the words of the prophets, apostles, and the Savior Himself. We can, and should, benefit from science. We should also remember that science is a human, and therefore flawed thing. While science may change its opinion, and consensus may be formed, truth does not vary. Sometimes science just have to catch up with truth.

---

[150] In Conference Report, Oct. 1952, 51

The Wave, Ashley Annajean Miller

SYMBOLS, SHADOWS AND STONE

When reading stories in scripture, we often find ourselves trying to decide: is this story just symbolic, is it a retelling of another story we've seen elsewhere, or is it fact? In short; is it a symbol, a shadow, or stone? It's our nature to attempt to classify things. We want things to fit into neat compartments. It makes it easier for us to process them.

For instance, if something is symbolic, we are not necessarily required to define an historical setting, or to identify the characters in the story to any great degree. We become more artistic in our thinking. We begin to look for the parables, the greater lessons to be learned. They are just stories, although perhaps important ones.

We use symbols in our lives every day. We don't have to get close enough to a stop sign to read the four characters it contains to slow down and prepare to stop. We recognize the shape and color, and we react accordingly. It is a symbol with meaning.

Many symbols are much more ancient. We should recognize a five-pointed star as a symbol for Christ. It has been so since eternity, and only in the nineteenth century did confusion arise.

When it appears on a temple, as in the Nauvoo Temple windows, it still retains this symbolism, although it also has physical form. As we see in this volume, the rainbow is a symbol with deep significance. It reminds us of the promise, or covenant, the Lord made with Noah, and that is inherited by us.

If it is a shadow of some other event, our thinking must deepen. We now begin to look for parallels. Who do these characters represent, is there a Christ figure? Is there a "man" figure, or perhaps an "Israel" figure. The scriptures are replete with examples. We often consider David a Christ figure, His shadow pervading all of scripture, if we are looking. This is true.

We see David as a type of Christ in three significant ways. David was a shepherd, he tended his father's flocks. Christ is called "the good shepherd," ready to lay down his life for his sheep.

SYMBOLS, SHADOWS AND STONE

David was also a deliverer. He delivered his people by overcoming Goliath and delivering the Israelites from the Philistines. Christ delivered us from the results of sin.

David was a king. He gathered the tribes of Israel, and unified them. Through love and loyalty, he brought a people together. Through his love Christ will gather all of his children, and is Lord and King.

So, these indications that David was a shadow of Christ, may seem like a reason to assume that David was simply a literary figure, that telling his story was another way of reinforcing the story of Christ in this ancient book.

However, in 1993 a fragment of engraved stone was found in Northern Israel[151]. It was learned that this stone, called the Tel Dan Stela, contained a reference to the "House of David" (Beit David). It commemorates the victory of an Aramean king over two southern neighbors, the "king of Israel" and the "king of the House of David." It was engraved in the ninth-century B.C.

So, King David was a historical figure, and he was also a shadow or type of Christ.

---

[151] Ten Top Discoveries, *Biblical Archaeology Review*, July/August September/October 2009

Figure 34: Tel Dan Stela. Photo: The Israel Museum, Jerusalem/Israel Antiquities Authority (photograph by Meidad Suchowolski)

The Book of Mormon has been subject to many of the same archeological questions. Where are the sites identified in its pages? "In one instance...Nephi does preserve a local name, that of Nahom, the burial place of Ishmael, his father-in-law. Nephi writes in the passive, "the place which was called Nahom," clearly indicating that local people had already named the place. That this area lay in southern Arabia has been certified by recent Journal publications that have featured three inscribed limestone altars discovered by a German archaeological team in the ruined temple of Bar'an in Marib,

Yemen[152]. Here a person finds the tribal name NHM noted on all three altars, which were donated by a certain "Bicathar, son of Sawâd, son of Nawcân, the Nihmite." (In Semitic languages, one deals with consonants rather than vowels, in this case NHM.)

Figure 35: Altar That Contains "NHM" Reference

Such discoveries demonstrate as firmly as possible by archaeological means the existence of the tribal name NHM in that part of Arabia in the seventh and sixth centuries BC, the general dates assigned to the carving of the altars by the

---

[152] S. Kent Brown, "'The Place Which Was Called Nahom': New Light from Ancient Yemen," *Journal of Book of Mormon Studies* 8/1 (1999): 66-68; and Warren P. Aston, "Newly Found Altars from Nahom," *Journal of Book of Mormon Studies* 10/2 (2001): 56-61

excavators.[153] In the view of one recent commentator, the discovery of the altars amounts to "the first actual archaeological evidence for the historicity of the Book of Mormon."[154, 155]

Which leads us to stone. This refers to the stories we have decided actually happened. It is easier to believe the Doctrine & Covenants relates to actual people. I like to refer to the characters in the Doctrine & Covenants as people with "last names." We can relate to them, we can even visit the homes they built, and read the letters they wrote. It is only somewhat more difficult with the Book of Mormon. Most members of the Church, if asked, will state that they do believe that a man named Nephi actually lived, and that Lehi, a prophet, was his father. They may debate the location of his "promised land," but that is only fueled by their belief in the historical accuracy of the Book or Mormon.

Things begin to get more complicated when we introduce the Bible. The New Testament is given much of the same historical

---

[153] Burkhard Vogt, "Les temples de Ma'rib," in Y émen: au pays de la reine de Saba (Paris: Flammarion, 1997), 144; see also the preliminary report by Burkhard Vogt et al., "Arsh Bilqis"—*Der Temple des Almaqah von Bar'an in Marib* (Sana'a, Yemen, 2000)

[154] Terryl L. Givens, *By the Hand of Mormon: The American Scripture That Launched a New World Religion*, New York: Oxford Univ. Press, 2002, 120

[155] S. Kent Brown, "Nahom and the 'Eastward' Turn," *Journal of Book of Mormon Studies* 12:1 (2003)

relevance as the Book of Mormon (perhaps with the exception of the Book of Revelations), but the Old Testament is seen as a mine field. It's hard to say where the line gets crossed. Most of us believe that David actually lived, and was a king of Israel (although some people may not be sure about the story of Goliath). And Elijah and Isaiah's writings may be confusing, but few actually doubt they wrote them. But mention Moses, and the fractures begin to show. An Israelite baby placed in a woven ark and adopted by Pharoah's daughter? This is strange considering the reality of the Tabernacle in the desert to most people and all the effort expended trying to locate the Ark of the Covenant.

By the time we get to Noah, the crack has grown into a chasm, with many people doubting the literal nature of scripture. And going back further, the rest of the people "without last names" become less solid. Stone no more, we begin to look for the shadows, or as a last hope, the symbols.

You can see how problematic this can be, and why it creates such uncertainty in many "critical thinking" young people today. If there is no clear line where people become legends, for that is what much of the world has made of them, their power and relevance is diminished at least, and is on its way to being extinguished.

I propose here an alternate view. We do not need to place scripture into one of the three boxes. Rather, it imbues each of them with great power. To say something is stone, does not restrict its symbolic significance. And in identifying the types and shadows, we do not rob anyone of their historical relevance. Just the opposite, an incredible richness will begin to inhabit these stories.

This principle can have great impact on each of us. As we begin to apply the concepts of Symbol, Shadow and Stone to our own lives, we begin to recognize that they are very real and living things. What symbols have personal meaning for you? I am a real fan of symbols. I like to surround myself with them. Some are the ancient and religious ones, the symbols that the Lord has defined since before time began. These figure prominently in our home (I have a very supportive wife). But I also appreciate modern symbols, with specific and personal meaning for me. I place them throughout my home. The screen on my smart phone contains a photograph that is symbolic to me. When I see it, it doesn't simply represent an image, but an experience and a feeling. Understand that the symbols you surround yourself with carry meaning, it is inherent in them. Consider it seriously. What do these symbols convey to you, and to your family each day?

How does the concept of shadows affect us? I think it is interesting that when we study scripture, it is so easy for us to

SYMBOLS, SHADOWS AND STONE

begin to recognize those who became Christ figures, shadows of the Savior. Through the sacrifices they made, the lives they lived and the service they offered, we begin to see His presence, His shadow. As each of us operate under the influence of the Light of Christ, and tune ourselves more and more to it, we can become shadows of Christ as well. Most assuredly for our families, and those who follow us, but also to our friends and communities.

I was recently told a story, in it there were two brothers, who lived in a small community. The brothers were struggling to make ends meet, and began stealing sheep as a way to survive. It wasn't long before the local villages noticed their missing sheep, and the culprits were identified.

The penalty for steeling sheep in this village was to have two letters branded into the forehead of the offender to mark them as a Sheep Thief (ST). An ST was placed on their forehead to remind them and others of what they had done. One brother couldn't bear the idea of being branded for life as a thief, and so he fled town. The other brother recognized his mistake, and decided he would do all he could to make amends to those he had wronged. He dedicated his life to the service of the people of his little village.

Years passed, and the man who stayed in the village grew old in the service of the members of his community. Eventually his body became bent, wrinkles appeared on his face, but the brand never faded. One day a man came to town and noticed the old man, and the mark on his forehead. He asked a young boy about the man. He mentioned the mark on his forehead, and asked about its meaning. The young boy thought for a moment and replied "I'm not really sure, it's been a long time. But I think it's an abbreviation for Saint."

The old man's behavior, in spite of his past, had made him a shadow of the Savior. As we act in accordance with the light within each of us, His light, we begin to look more and more like Him, until at last, we are a shadow of Him.

And finally, Stone. These are the moments when we get to see what is actually happening. The world creates such illusions. They have all the trappings of reality, and yet they are fleeting and insubstantial. There have been moments in my life when I have been able to glimpse stone, bedrock even. I have witnessed the real workings of the Lord in my life, I have seen behind the curtain, and in that fleeting glimpse have seen the great plan He has for me. Years ago there was a song that contained the lyric "he carried in his heart a picture of the man he knew he would become." I have reflected on these words many times in my life. Heavenly Father gives us these pictures, these glimpses, as a result of our faith, but also to build our faith, and to help us move forward.

None of these principles is more important than the others, and it is my hope that we will experience scripture in a new way as we begin to allow them to exist alongside each other. I hope the same is true in our lives as well, allow Symbols, Shadows and Stone to become a conscious part of your life.

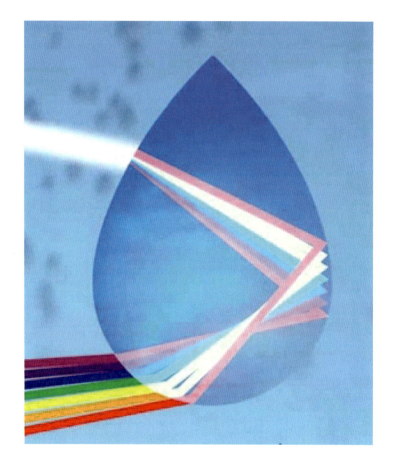

LET THERE BE LIGHT

Rainbows are objects of light. They occupy no physical space, and have no substance that can be quantified by any means available to us. Our discussion of the token of the bow, must be based on a clear understanding of light, and the challenges it presents.

A review of the creation account in the Book of Genesis reveals an interesting piece of critical information. Much of what will be discussed in this chapter, is based on the insightful chronology established in Moses' account.

In the Lectures of Faith, Joseph Smith taught:

> "When a man works by faith he works by mental exertion instead of physical force. *It is by words*, instead

of exerting his physical powers, with which every being works when he works by faith. God said, 'Let there be light: and there was light.' ... And the Saviour says: 'If you have faith as a grain of mustard seed, say to this mountain, "Remove," and it will remove; or say to that sycamore tree, "Be ye plucked up, and planted in the midst of the sea," and it shall obey you.' Faith, then, works by words; and with these its mightiest works have been, and will be, performed. ...

"... The whole visible creation, as it now exists, is the *effect of faith*. It was faith by which it was framed, and it is by the power of faith that it continues in its organized form, and by which the planets move round their orbits and sparkle forth their glory".[156]

It is interesting that the act of creation was therefore an act of words, which is the effect of faith. The Lord said, "let there be light, and there was light." No other object was required. Not a sun (not created yet), not any other star (not created yet), just a word, or extending Joseph's thoughts, just the faith of the Savior. Light existed before the creation of the sources the world looks to for light. Light is independent of them.

---

[156] Lectures on Faith, 72–73

Elder John Taylor went further when he explained that God "caused light to shine...before the sun appeared in the firmament; for God is light, and in him there is no darkness. He is the light of the sun and the power thereof by which it was made; he is also the light of the moon and the power by which it was made; he is the light of the stars and the power by which they are made"[157]

I believe that the light we observe is a fractional light, just a glimpse into the light brought into the universe by the faith of the Savior. It is interesting that most science textbooks refer to light as both a particle and a wave. They call this the "dual nature of light." This refers to its behavior as something with mass and momentum that can exert force, and something without mass that can travel through the void of a vacuum.

Consider Joseph Smith's comments on matter:

> "There is no such thing as immaterial matter. All spirit is matter, but it is more fine or pure, and can only be discerned by purer eyes; we cannot see it; but when our bodies are purified we shall see that it is all matter"[158]

---

[157] Journal of Discourses, 18:327
[158] Doctrine & Covenants 131:7-8

The idea of a dual nature for light is supported by both Joseph Smith's comment above and Parley Pratt's statement below:

"Matter and spirit are the two great principles of all existence. Everything animate and inanimate is composed of one or the other, or both of these eternal principles.... Matter and spirit are of equal duration; both are self-existent, they never began to exist, and they never can be annihilated.... Matter as well as spirit is eternal, uncreated, self existing. However infinite the variety of its changes, forms and shapes; ...eternity is inscribed in indelible characters on every particle"[159]

The two characteristics of light are not mutually exclusive as once thought. For many decades it was assumed that light could only be observed behaving as one or the other, particle or wave. But, in 2015 a researcher at École Polytechnique Fédérale de Lausanne (EPFL), Fabrizio Carbone, provided the first-ever evidence of light as both a particle and a wave (Figure Below). His photograph is reminiscent of the focus of this book, rainbows, displaying the variety of wavelengths of light created as electrons flowed through a silver wire.[160]

---

[159] History of the Church 4:55
[160] "Simultaneous observation of the Quantization and the Interference Pattern of a Plasmonic Near-field", *Nature Communications* 6, Article 6407

SYMBOLS, SHADOWS AND STONE 183

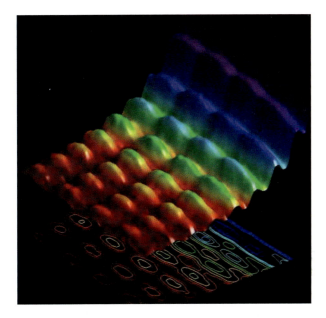

Figure 36: Light as a Particle and a Wave

Light is not either/or, it is both, simultaneously. It should be no surprise to any of us that believe in this dual nature of matter (physical and spiritual) to accept a dual nature for light (physical and wave energy). Rainbows may be a way for us to begin to learn more about the "true" nature of light.

This dual nature, as particle and wave, also extends to the cause for its behavior when reflected and refracted. The reflection of light (as off the inside of the raindrop as we will see in the next chapter) is best explained when light is considered as a particle. When light is reflected from a smooth surface, both approaches (particle and wave) seem to describe observations accurately, but when the reflective surface is rough, light appears to behave more in accordance with what would be predicted by a particle.

With refraction (as light moves from one medium to another) however, the roles are reversed. The wave nature of light seems to describe the observed behavior more accurately. Those who suggest a particle nature for light in refraction require the acceptance of a "special force" to cause refraction to occur. Both of these behaviors, reflection and refraction, are factors in the formation of a rainbow as we will see.

Light energy is the most basic energy of all, so it is not surprising that the first words spoken by God in Genesis were: "Let there be light".[161] In fact, the division created between light and darkness was not related to the creation of suns or stars either, since soon after declaring the light "good", he divides it from the darkness, and calls light "Day" and darkness "Night". In Hebrew, day is Yom (יום), it refers not only to a twenty-four hour period, but also to a "period of light". It also can refer to a time period of light without specific duration.

The Hebrew word for "darkness," araphel (ערפל), can be derived from the root araph (ערף) meaning "to drip." Another noun derived from this root is ariph (עריף) meaning "cloud" (as being the source for "rain"). Furthermore, ערפל would then literally

---

[161] Genesis 1:3

mean "thick cloud," thereby meaning "darkness" (as in blocking the sun).

It was not until 2 days later, that the Sun and the Moon and the stars were created. And all of these lights—and the light which they generated and sent forth to be "for signs, and for seasons, and for days, and years"[162] would also serve life itself through the many marvelous processes it would fuel for Earth's coming inhabitants.

Thus it is that: "In Him [that is, the Word of God] was life; and the life was the light of men"[163]. It then follows also that Christ is "the true Light, which lighteth every man that cometh into the world"[164].

And that is true in both the physical sense and spiritual sense. Physically, "in Him we live, and move, and have our being" so that He is "not far from every one of us"[165]. It should be painfully sobering for even those who refuse to believe in Him to realize suddenly (as they must, someday) that their very existence, even the cellular structure of their bodies, depends on His moment-by-moment maintenance. If He just withdrew His power for an instant we would collapse into nothingness.

---

[162] Genesis 1:14
[163] John 1:4
[164] John 1:9
[165] Acts 17:28,27

Spiritually, we are likewise assured that He enlightens "every man that cometh into the world"[166]. That is, even those born in some remote home and those who may spend all their lives without ever hearing of Christ, have been given some spiritual light so that if they respond positively to the light they have, will then somehow be given more and more light, eventually enough to be led to Him.

So much of scientific thought, particularly in the area of Physics, is tied to the idea of light. The most famous scientific formula in history depends on its infallibility:

$$E=mc^2$$

It has been assumed that light, as we observe it, is the way light has always been. Physics of course does not concern itself with the fact that light predates suns, stars or physics. It simply assumes that light has always been what it is, and then creates theories, that become laws. Who is supposed to obey these laws?

In violation of these "laws", in September 2011, a group of scientists announced that they had detected subatomic particles travelling at speeds greater than the speed of light in vacuum. This finding appears in every way to be in conflict with Einstein's theory of relativity.

---

[166] John 1:9

Figure 37: Albert Einstein, 1921

And there are physicists now who are beginning to question the infallibility of light, in particular its absolute speed. They believe that in the early universe, it may have travelled much faster than it does today. In fact, the physicists responsible for this new theory propose that at the time of creation, light may have had infinite speed. That is, it knew no limitation.

They say of their ideas, "In our theory, if you go back to the early universe, there's a temperature when everything becomes faster. The speed of light goes to infinity and propagates much faster than gravity..."[167]

The obvious conclusion is that science today still has a lot to learn about light. And, so do we, both in terms of science, and faith. True light is the divine energy, power and influence that gives life and light to all things. As scientists are beginning to remember, the Light of Christ "proceedeth forth from the presence of God to fill the immensity of space." It is "the light which is in all things, which giveth life to all things, which is the law by which all things are governed".[168] It is eternal, and infinite.

A subsequent chapter on the science of rainbows will put this information into practice.

---

[167] Niayesh Afshordi and Joao Magueijo, *Critical Geometry of a Thermal Big Bang*, Physical Review, 18 November 2016
[168] D&C 88:12-13

WHY ARE THERE SO MANY SONGS ABOUT RAINBOWS?

You can't really write a book about rainbows, and not mention the legendary scholar, and green frog, Kermit. This song is part of our past (for those of my generation), and few can resist humming along:

> Why are there so many
> Songs about rainbows
> And what's on the other side
> Rainbow's are visions
> They're only illusions
> And rainbows have nothing to hide"

In the late 1970's and early 1980's, "The Rainbow Connection" brought rainbows into the common dialog, as *The Wizard of Oz* and "Somewhere Over the Rainbow" had done forty years earlier. Both songs portray rainbows in an ethereal, hopeful, and longing light (no pun intended). This kind of entertainment is largely responsible for common culture's relationship with rainbows.

You may not be aware that "Somewhere Over the Rainbow" was written, not about the mythical Land of Oz, but about Israel, the homeland of the Jews.

The lyrics to this beloved song were written by Yip Harburg, the youngest of four children born to Russian-Jewish immigrants. His real name was Isidore Hochberg, and he grew up in a Yiddish-speaking, Orthodox Jewish home in New York.

The song's music was written by Harold Arlen, also a cantor's son. His real name was Hyman Arluck, and his parents were from Lithuania.

Together, Hochberg and Arluck wrote "Somewhere Over the Rainbow," which was voted the 20th century's No. 1 song by the Recording Industry Association of America and the National Endowment for the Arts.

In writing it, the two men reached deep into their immigrant Jewish consciousness (imbedded by the pogroms of the past and the Holocaust that was about to happen) and wrote a timeless melody set to the prophetic words: "And the dreams that you dare to dream, really do come true."

But man's history with rainbows is ancient. The first scientific analysis of rainbows is attributed to the Greek scholar Aristotle. And, "Despite its many flaws and its appeal to Pythagorean numerology, Aristotle's qualitative explanation showed an inventiveness and relative consistency that was unmatched for centuries. After Aristotle's death, much rainbow theory consisted of reaction to his work, although not all of this was uncritical."[169]

Figure 38: "Rainbow Clouds" (Jackson Obrien, 2018)

---

[169] Raymond L. Lee and Alistair B. Fraser (2001), *The Rainbow Bridge: Rainbows in Art, Myth, and Science*, Penn State Press, 109

It is interesting that in Jewish tradition, a rainbow marks a people that are sinning. A generation without a rainbow is seen as one with an especially high level of spirituality. There is disagreement on this among Jewish scholars though, the Midrash and some Rabbi's disagreeing with this interpretation.

The Prophet Joseph himself said "whenever you see the bow withdrawn, it shall be a token that there shall be famine, pestilence, and great distress among the nations, and that the coming of the Messiah is not far distant." This implies that rather than signifying a time of great spirituality, the absence of the bow is a sign of the return of the Savior, the Lord himself saying: ""In the days of Noah I set a bow in the heavens as a sign and token that in any year that the bow should be seen the Lord would not come..."

The prophet Ezekiel described a vision in which he had seen the divine presence "As the appearance of the bow that is in the cloud in the day of rain, so was the appearance of the brightness round about. This was the appearance of the likeness of the glory of the LORD. And when I saw it, I fell upon my face, and I heard a voice of one that spake. ...."[170] This experience brings to my mind the discussion on light, and its divine origin and power, discussed in the previous chapter.

---

[170] Ezekiel 1:28

Because of this vision, there was a Jewish sage that said that when a person sees a rainbow, he should bow down, prostrating himself in front of God. Others, however, said it was forbidden to do so because it would look like one was worshipping the rainbow.

Since the rainbow is said to represent the beauty of the divine presence and the glory of God, those who follow the teachings of the Talmud believe that it's not proper to stare at a rainbow.

The optics of rainbows was investigated by Theodoric of Frieberg, a Dominican monk from Germany. In 1304 he began his work and in 1310 completed a manuscript that identified the reflection and refraction process inherent in the creation of a rainbow. Reflection occurs when light changes its direction in a single medium (such as inside the raindrop). Refraction occurs when light changes its direction as it passes from one medium to another (such as on the surface of the raindrop as the water enters and exits it). Theodoric concluded that, as a result of the process of refraction and reflection, one drop of water would send only one color of light to the eye of the observer. The rainbow results from a combination of many drops of water in a cloud at different positions from a viewer, where the drops at each distance from the viewer send a particular color of the rainbow.

The first rainbow calculations were carried out by René Descartes back in 1637. Isaac Newton is often credited with using a prism to split "white" light into its component wavelengths, however, the truth is even more interesting. Not only did he use a prism to break the light into a multitude of colors in 1665, but he also believed that all the colors he saw were in the sunlight shining into his room. As a result, he thought he should be able to combine the colors of the spectrum and make the light white again. To test this, he placed another prism upside-down in front of the first prism. He was right. The band of colors combined again into white sunlight. Isaac Newton was the first to prove that white light is made up of all the colors that we can see. Prior to his experiment, there was not consensus that these different wavelengths represented all the components of white light.

Think about this for a moment. The light we see each day, radiating down upon us, is actually composed of an infinite number of varying wavelengths of radiation. What a prism (or a raindrop) does is cause each of those beams of light to divulge what they really are. Together they create a beautiful white light, but individually they are red, orange, yellow, green, blue, violet or somewhere in between. It is the process of reflection and refraction, and their nature, that allow us to see what they really are: beautiful, complex, varied, infinite, and precious.

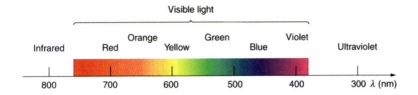

A rainbow is formed by sunlight streaming from behind us, internally reflected back towards us by a multitude of tiny suspended raindrops. When light meets an interface between air and water (the surface of a raindrop) some of it is reflected and some of it passes through. The angle of refraction depends both on the light's wavelength and on the angle at which it hits the surface. Parallel rays entering a spherical raindrop bounce off the inside and come back towards us. The reflection angle from the back of the raindrop, combined with the two refractions (one as it enters the drop, and one as it leaves) conspire to concentrate each wavelength of light at a certain return angle. Wavelength corresponds to color, so the colors separate into the familiar bands.

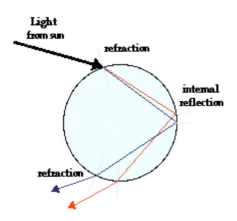

Rainbows are only formed when specific criteria are met. The sun must be behind us, raindrops in front of us, and must be unobstructed. Morning and later afternoon are the best times to see a rainbow due to the angle of the sun in relation to the raindrops.

Consider the parallels. The light which creates this token must be behind us (and radiating over us) with nothing between us and it. The token will only be revealed when the storm has past (or is in the distance). And, in spite of the storms of the night, the morning can bring the promise. Or, if the day is full of clouds...wait, the bow may still appear.

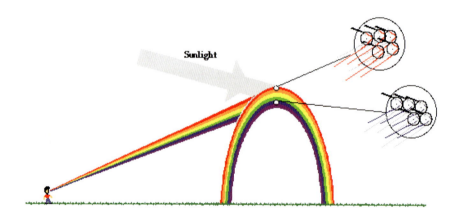

One factor that leads to interesting biblical interpretation is the dramatic effect of raindrop size. The best rainbows are narrow ones with intense colors and these are made by large raindrops. Such as during heavy rainstorms.

Smaller drops produce broader bows with less saturated colors. Very small drops give nearly colorless cloudbows and fogbows. So, one possible explanation of the bow's mention in the books of Moses, and the implication that it is something new, could very well be the nature of rain before and after the flood.

Figure 39: A Cloud Bow

If the majority of rain prior to the flood was of the "tal" variety, a light rain, or dew, while those after the flood were of the "geshem" variety, or heavy rain or the "geshem shotef" of the torrential rain (as in Ezekiel) we would see almost colorless rainbows prior to the flood, and vibrant, intense rainbows afterwards. All based on the size of the raindrops.

In a conversation with Temple muralist Linda Curley Christensen, the following wonderful experience was related to me:

During the construction of the Nauvoo Temple, numerous artists were asked to paint murals to adorn the rooms of the Temple. Linda was already painting a mural for another Temple, so when the time came for Church leaders to review the work, she was asked to bring her mural to the BYU Motion Picture Studios and include it in their review. She brought her mural to the Studios, and when President Hinckley arrived, he began to walk past the numerous murals. At one point, he paused at the mural of Robert Marshall. He noted a beautiful rainbow that had been placed above a waterfall. He smiled, commenting that he liked it. But, he seemed to hesitate as well, and said he would speak with the artist again about it. It was several days before the call came, and he asked Robert Marshall to remove the rainbow from the mural, stating that there had been no rainbows at that point, prior to the flood.

Our scientific understanding of rainbows may seem very robust now, but it is not complete. There are undoubtedly details that can still be improved with additional study, and with the Lord's help.

Scientists must avoid falling into the trap of defending all aspects of current thought just because they feel the underlying "truth" needs protecting. The most effective way to find a deeper description of nature is to seek the spirit, study the eyewitness accounts in scripture, and perform more observations. If necessary, each of these efforts may push the existing conception to the breaking point, until truth is revealed.

And so, we're back to our friend Kermit. He understood the nature of rainbows, perhaps better then we imagined. He said; "Rainbows are visions," and although he continued "they're only illusions" he did conclude that, "someday we'll find it, the rainbow connection. The lovers, the dreamers, and me."

Rainbow Cake (Violet Obrien, 2018)

IS IT REALLY A RAINBOW AFTER ALL?

Any symbol is vulnerable to attack. It may be assumed by another because of graphic appeal. Or, it may be deliberately plagiarized to create confusion.

With the recent (within my lifetime) promotion of the rainbow as a symbol of something other than what it has meant for millennia, it is fair to ask if the flag being paraded about (literally in many cases) is actually intended to be a rainbow. Obviously, just placing colored stripes alongside each other does not necessarily mean that the intention is to connote a rainbow. But, this approach is used in art, children's books, and displays worldwide, with the intention that it convey the symbol of the rainbow.

It really comes down to the intentions of the creator, and what was intentionally being conveyed. The rainbow currently being used to portray gay pride actually began in the 1970's, in San Francisco. A man named Gilbert Baker created the first flag (originally with eight colors: hot pink (representing sexuality), red (representing life), orange (representing healing), yellow (representing the sun), green (representing nature), turquoise blue (representing art), indigo (representing harmony) and violet (representing spirit). Only a few of these flags were made, and were flown in the 1978 "Gay Freedom Day" Parade in San Francisco. Baker described it as "the Rainbow Flag."

A manufacturer was approached to mass produce the flag, but the hot pink fabric needed was not available, so a seven-color flag was utilized (red, orange, yellow, green, blue, indigo and violet). Ironically, these were surplus flags that had already been made for the International Order of Rainbow for Girls, a Masonic organization for young women.[171]

---

[171] "A Brief History of the Rainbow Flag," *San Francisco Travel Magazine*, 2018

IS IT REALLY A RAINBOW AFTER ALL?

The published mission of this group is interesting:

"Rainbow is a nonprofit organization that strives to give girls the tools, training, and encouragement to let their individual spirits shine bright. By providing members with a safe, fun, caring environment where responsible, older girls can interact and mentor younger girls through family involvement."[172]

Regardless of their original purpose, "to mentor younger girls through family involvement", this flag (and symbol) had now been assumed by a group with another agenda.

It is also interesting to consider that eight colors comprised the original gay pride flag, in counter-point to the eight values of the Young Women in the Church. Equally interesting is the fact that there were originally seven, until 2008 when the eighth

---

[172] The International Order of the Rainbow for Girls

value was added: Virtue. The Values, and ultimately the colors used to represent them graphically, trace their origins to the early 1900's. It was the Beehive Handbook that outlined the first recognition program for young girls. It included seven "fields" of personal improvement: Religion, Home, Health, Domestic Arts, Out of Doors, Business and Personal Service.

In 1979, the San Francisco LGBT Pride Celebration Committee split the colors into two flags, eliminating the indigo stripe, to allow it to be flown in two separate sections with three colors each, on opposite sides of the street for the 1979 Gay Freedom Day Parade. This six-stripe version is the one most commonly seen today.

It is clear then that the creator of the flag so prominent in the Gay community did intend to create a rainbow flag. What does that mean? Not much actually, since calling something by a given name doesn't actually make it so. A dog is a dog, even if you call it a cat. However, it does lead to the question of motive. That does matter. Knowing not just how, but why this alternate use has been presented can protect us. It is not a matter of confusion, it is a matter of misrepresentation, and deception.

As I discussed in the chapter "A Flood, a City and a Garden" the rainbow did not arise out of just any circumstance, but as a result of the great flood that destroyed most of humankind. The events which precipitated this flood were directly related to family. It was the loss of Noah's granddaughters to the evil influences of the world, the fact that people "hated their own blood," and that children were being born into a world where their agency was denied them.

The flood was directly related to families, and the attack on them. Who originated this attack? The same force and being that has attacked this foundational element of Heavenly Father's plan since before time began. Satan understands the significance of family more than most of us. He has continually attacked it, undermined it, destroyed it, and now is trying to rename it.

A prominent columnist wrote of our day: "One thing is certain. We shall be given no centuries for a leisurely and comfortable decay. We have an enemy now—remorseless, crude, brutal and cocky ... [who believes] that we are in an advanced state of moral decline ... [and] ripening for the kill."[173]

The rainbow is not simply a colorful symbol, it is a token given to us, to our families. It was the attack on family that caused the flood, and it was the hope for families that gave us the rainbow.

---

[173] Jenkin Lloyd Jones, *Human Events*, November 24, 1961

THIS SHALL BE A TOKEN

In the Old Testament, the proximity of two references to the term "token" is important. Not only does it occur in Genesis 9:13, (לְאוֹת) but also in Exodus 12:13 (לְאֹת). It is interesting because in the first instance God is referring to the token given to Noah; "I do set my bow in the cloud, and it shall be for a **token** of a covenant between me and the earth." And in the second it refers to the manner the Israelites would use to separate themselves from the Egyptians, "And the blood shall be to you for a **token** upon the houses where ye are: and when I see the blood, I will pass over you, and the plague shall not be upon you to destroy you, when I smite the land of Egypt."

It is clear by these two references, that simply interchanging token and symbol does not suffice. With a token, a physical action is required.

Further emphasis is placed on this use of language by the covenant it is associated with. The Hebrew word for covenant is berith (בְּרִית). It means to cut or divide, and refers to the sacrificial custom in covenant making between two parties where an animal is divided in two halves. The parties of the covenant then walk between the pieces to establish the covenant.

In Genesis 15, the Lord enters into a covenant with Abram regarding his inheritance of the land. He instructs Abram to prepare a heifer, a she goat, a ram, a turtledove and a young pigeon in this manner. In this way "the LORD made a covenant with Abram, saying, Unto thy seed have I given this land, from the river of Egypt unto the great river, the river Euphrates…" (Genesis 15:18) In Genesis 17, the Lord renews this covenant with Abram (in Hebrew; My Father above is exalted), and changes his name to Abraham (in Hebrew; Father, by grace of God, of a multitude).

We see a further example of the relationship of token and covenant in Alma 46:21. After Moroni has raised the Title of Liberty the people came running "…rending their garments in token, or as a covenant, that they would not forsake the Lord their God."

It is significant that not only did the people "rend their garments", dividing them as an act of covenant, but Moroni had as well. He "rent his coat; and he took a piece thereof and wrote upon it —In memory of our God, our religion, and freedom, and our peace, our wives, and our children..."[174] His action represented the covenant he was making with God and with the people.

These tokens are intimately tied to the covenants being made. They are a physical act, that typifies the covenant. They are related.

This concept of the physical action related to the covenants, and ordinances, inherent in this process is further exemplified in the ordinances we participate in within the walls of the Temple. Brigham Young expressed it this way:

> "Your endowment is, to receive all those ordinances in the House of the Lord, which are necessary for you, after you have departed this life, to enable you to walk back to the presence of the Father, passing the angels who stand as sentinels, being enabled to give them the key words, the signs and tokens, pertaining to the Holy Priesthood,

---

[174] Alma 46:12

and gain your eternal exaltation in spite of earth and hell"[175]

It is the understanding of these physical tokens, and the covenants they are integral to, that will identify us as one of the elect of God.

So, what was the token offered in the bow, and what action was taken? The "token of the bow" was an actual, physical act on the part of a loving Heavenly Father to represent the covenant he made with Noah, his family, every generation that would follow, and all living creatures. This covenant will be reviewed in a subsequent chapter, but the token associated with it requires a physical act, not just at the time Noah saw it, but each time it is painted in the heavens. Each and every time a rainbow appears, this same light, created by the Savior, is refracted and reflected, exhibiting its dual nature, millions and millions of times to produce this token. With the sun and its glory behind us, we are prepared to witness this amazing phenomenon, after the storm. It is, and will be, a tangible representation to everyone who remembers.

---

[175] Journal of Discourses, 2:31

THE COVENANT

Most books about rainbows and the flood are understandably preoccupied with Noah. The story seems to be about Noah, and his family, and all those animals. But, as you have undoubtedly noticed, this book spends significant time discussing his great-grandfather, Enoch. In this chapter, I hope the reason for this will become more clear. Not only was there a familial relationship between Enoch and Noah, but there was also a covenantal relationship between these two great Prophets.

As we have seen throughout this volume, the rainbow is a token of a covenant. We have to be grateful once again to the Prophet Joseph Smith for giving us restored scripture that identifies the full ramifications of this covenant. In the Old Testament, we read that the bow "shall be in the cloud; and I will look upon it, that I may remember the everlasting covenant between God and every living creature of all flesh that is upon the earth."[176]

The covenant described here states that "neither shall all flesh be cut off any more by the waters of a flood; neither shall there any more be a flood to destroy the earth."[177]

The covenant described in Genesis is between Noah, and his descendants, and every living creature, that he will never again destroy them or the earth by sending a flood. It's powerful, hopeful, and as we have described, meaningful for our day.

But as beloved storyteller Paul Harvey used to say, now for "the rest of the story." In what we call the Joseph Smith Translation (JST), and what Joseph himself simply called a "new translation," we find much additional light and knowledge about the nature of this far-reaching covenant.

The Joseph Smith Translation of Genesis 6:18 reveals one major benefit of this new revealed text. In it the Lord told Noah, "With thee will I establish my covenant, even as I have sworn unto they father, Enoch, that of thy posterity shall come all nations."

So we see that one characteristic of this covenant is that it was first made to Enoch. In the Book of Moses, the Lord gives

---

[176] Genesis 9:16
[177] Genesis 9:11

THE COVENANT

Enoch a great vision of things to come, including his descendant, Noah. When this vision of the wickedness of man had overwhelmed Enoch, so much so, that he refuses to be comforted, the Lord shows him the descendants of Noah, and the coming of the Son of Man. The Lord shows Enoch that it will be through his family that Noah will come, and through Noah, ultimately the Savior.

After seeing all that would come to his descendants, Enoch asks the Lord, "O Lord, in the name of thine Only Begotten, even Jesus Christ, that thou wilt have mercy upon Noah and his seed, that the earth might never more be covered by the floods."[178] Because of Enoch's righteous desire, the Lord says he cannot deny the request and, "He covenanted with Enoch, and sware unto him with an oath, that he would stay the floods; that he would call upon the children of Noah; And he sent forth an unalterable decree, that a remnant of his seed should always be found among all nations, while the earth should stand..."[179] From this scripture we learn that it was Enoch's request that brought the covenant, and with it the token.

Through the restoration of Moses' record, we also learn much more about what the covenant actually promises. It is so significant, that it is quoted here in its complete form:

---

[178] Moses 7:50
[179] Moses 7:51-52

> "And the bow shall be in the cloud; and I will look upon it, that I may remember the everlasting covenant, *which I made unto thy father Enoch; that, when men should keep all my commandments, Zion should again come on the earth, the city of Enoch which I have caught up unto myself. And this is mine everlasting covenant, that when thy posterity shall embrace the truth, and look upward, then shall Zion look downward, and all the heavens shall shake with gladness, and the earth shall tremble with joy; And the general assembly of the church of the firstborn shall come down out of heaven, and possess the earth, and shall have place until the end come. And this is mine everlasting covenant, which I made with thy father Enoch. And the bow shall be in the cloud, and I will establish my covenant unto thee, which I have made* between *me* and *thee, for* every living creature of all flesh that *shall be* upon the earth. And God said unto Noah, This is the token of the covenant which I have established between me and *thee; for all* flesh that *shall be* upon the earth."[180]

So, while the world looks to the rainbow as a reminder of the flood, and a promise that it will not come again, we can also look to the return of the City of Enoch and with it the "general

---

[180] JST Genesis 9:21-25, *italics represent new translation*

assembly of the church of the firstborn." This includes those who were residents of the City of Enoch, but also those who have been translated throughout history, and those who are resurrected at the coming of the Savior.

The description of this event in the Book of Moses is inspiring. As the City of Enoch joins the New Jerusalem, the reunion is a glorious one. The Lord Himself describes it this way:

> "Then shalt thou and all thy city meet them there, and we will receive them into our bosom, and they shall see us; and we will fall upon their necks, and they shall fall upon our necks, and we will kiss each other;

> "And there shall be mine abode, and it shall be Zion, which shall come forth out of all the creations which I have made; and for the space of a thousand years the earth shall rest."[181]

This is a reunion we all want to be a part of. And, just as he told Noah, "with thee will I establish my covenant, even as I have sworn unto they father, Enoch…" All of this comes to those who inherit the covenant of Noah (and of Enoch). The token of the bow is a token of this miraculous covenant, that the Lord will never again destroy the world by flood, that he will remember the posterity of Noah, and that one day the glorious City of Enoch will return, with all its righteous inhabitants and join the New Jerusalem, where Christ himself will join the Saints.

---

[181] Moses 7:63-64

THE COVENANT

Mark & Esther Stringer, 2018

THE TOKEN OF THE BOW

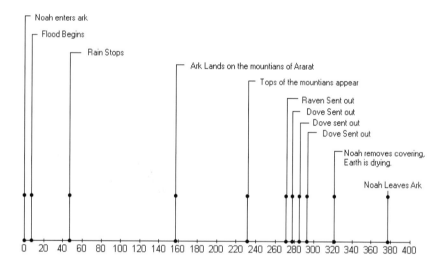

AFTER THE FLOOD

When the flood waters finally withdrew from the earth, Noah and his family left the ark. And all the animals that had been protected in the ark, were able to go into the world.

In leaving the ark, Noah teaches another lesson. In spite of all the challenges he had just endured, or perhaps because of them, Noah remembered the reason he had been protected. He remembered the Lord.

> "And Noah built an altar unto the Lord; and took of every clean beast, and of every clean fowl, and offered burnt offerings on the altar. And the Lord smelled a sweet savour; and the Lord said in his heart, I will not again curse the ground any more for man's sake; for the imagination of man's heart is evil from his youth;

neither will I again smite any more every thing living, as I have done. While the earth remaineth, seedtime and harvest, and cold and heat, and summer and winter, and day and night shall not cease."[182]

Upon leaving the ark, Noah offered sacrifices to the Lord. It was a significant sacrifice, one of every clean animal he had taken on the ark (of the seven clean animals who sought refuge in the ark before the flood). Scripture then says that the Lord smelled the sweet savour, and said in his heart that he would not again curse the ground for man's sake. This "sweet savour" is pleasing to the Lord because it represents the commitment of Noah to worship him, and recognize His hand in the saving Noah and his family. This kind of sacrifice is "sweet" and "pleasing" to the Lord. God accepted Noah's (and the earth's) sacrifice, and lifted the curse He placed on it when Adam fell.

Because Adam "...hast eaten of the fruit of the tree of which I commanded thee, saying—Thou shalt not eat of it, *cursed shall be the ground for thy sake*; in sorrow shalt thou eat of it all the days of thy life. *Thorns also, and thistles shall it bring forth* to thee, and *thou shalt eat the herb of the field*. By the sweat of thy face shalt thou eat bread, until thou shalt return unto the

---

[182] Genesis 8:20-22

ground—for thou shalt surely die—for out of it wast thou taken: for dust thou wast, and unto dust shalt thou return."[183]

As we saw in a previous chapter (*Noah*), there is a tradition that the world did not behave itself after the fall. It is certainly true that the world Adam and Eve were introduced to was a very foreign and frightening place. The animals that roamed the earth were violent (as seen by the fossil record), and through the scripture we just reviewed, thorns and thistles (weeds) were now prevalent.

The lifting of this curse with Noah is no small thing. We often think that the world we live in is the world Adam and Even would have experienced. We now know that the world we experience is a post-flood world that has had the curse of Adam lifted. While it may be difficult and trying, it is a world that is behaving itself, while the world Adam experienced after the fall did not.

There is one final element to the promise made to Noah after his sacrifice. God tells us that as long as the earth remaineth, there will be seasons and harvests. While the earth is recovering from the trauma of the flood, and the fountains of the deep coming forth, none the less, there will be seasons and harvests until He returns.

---

[183] Moses 4:23-25, Genesis 3:17-19 (Emphasis Added)

After the flood withdrew, the ocean's heat would take at least 600 years or so to dissipate, during this period the "Ice Age" dominated. Job lived soon after the Flood, and his book contains more references to ice and snow than the rest of the Bible put together.

Weather patterns after the Flood were very different from what Noah's family had known prior to the deluge. Widespread volcanic activity combined with warm ocean temperatures set the stage for the Ice Age that soon followed.

Figure 40: "Longneck Dinosaur," (Danny Semadini, 2018)

After the Flood, Noah stepped into a forbidding world. The animals and plants from the previous world were dead and buried under thousands of feet of sand and mud. New plants were struggling to reestablish themselves across the barren expanse of earth.

Meanwhile, the earth remained unstable. It would take centuries to quiet down. Supervolcanoes belched ash and death over vast regions, and superquakes rocked the earth. Somehow, the earth's climate had to transition from the warmest, wettest period in history, to the moderate weather we expect today. But it was a rocky transition. The volcanic activity during the Flood had left the oceans very warm, an estimated average 86°F, in contrast to 39°F today.

Warm oceans next to cold, barren continents was a recipe for violent storms. In the years after the Flood, "hypercanes," similar to Jupiter's Great Red Spot, persisted for decades. These storms drew water from the oceans and rapidly dumped it onto the land. The water quickly filled the depressions in the continents, and these temporary lakes burst through their barriers, cutting deep canyons in their wake, such as the Grand Canyon. The intense rains saturated the newly laid sediments, allowing groundwater to blast miles of caves in days.

As the oceans cooled, precipitation declined, and many of the world's forests dried and converted to grasslands. When the earth had cooled sufficiently, precipitation began to fall as snow and ice, especially in the world's mountain ranges, Antarctica, and northeastern North America. The ice built up rapidly, sometimes miles thick. This explains the Mammoth's found buried in ice. The change was so rapid, the animals did not have the opportunity to migrate. In North America this ice eventually surged under its own weight, spreading out and scraping the earth's surface, and then it melted suddenly.

As the ocean-cooling hypercanes dissipated, the pattern of air circulation changed for the entire planet. With this shift, deserts formed in belts around the world, about 30 degrees north and south of the equator.

As the animals left the Ark, they rapidly multiplied and spread over the earth. Within just a few years, animals had reached every continent, including Antarctica, which was still warm. Many animals, such as tortoises, traveled on huge mats of floating logs that circled the earth's oceans for centuries following the Flood.[184,185]

Much of our knowledge of these events comes to us from the Book of Genesis, and the Book of Moses. But, one of the most compelling passages regarding the flood comes to us from the book of Ether, in the Book of Mormon. This is the story of the Jaredites, who left the area of the Tower of Babel, after the Lord warned them of the coming confusion. This occurred in the

---

[184] Wise, K.P. and Croxton, M., Rafting: a post-Flood biogeographic dispersal mechanism; in: Ivey Jr., R.L. (Ed.), *Proceedings of the Fifth International Conference on Creationism*, technical symposium sessions, Creation Science Fellowship, Pittsburgh, PA, pp. 465–477, 2003

[185] Van Duzer, C., *Floating Islands: A Global Bibliography*, Cantor Press, Los Altos Hills, CA, 2004

generations that remembered the flood from their parents. As Moroni abridged the Jaredite record, here are his words:

> "For behold, they rejected all the words of Ether; for he truly told them of all things, from the beginning of man; and that *after the waters had receded from off the face of this land it became a choice land above all other lands*, a chosen land of the Lord; wherefore the Lord would have that all men should serve him who dwell upon the face thereof; And that it was the place of the New Jerusalem, which should come down out of heaven, and the holy sanctuary of the Lord."[186]

Notice that only after the waters receded from off the face of this land (North America) did it become a choice land above all other lands, the curse having now been lifted. While it may have been important prior to the flood, as part of the original land mass that was lifted from the oceans of creation, only now could it again be considered a 'choice land.' Perhaps this was because now all the wickedness had been scrubbed from its surface, but choice it was. And, it would be the home of the New Jerusalem, "which should come down out of heaven,"[187] Enoch's city, the fulfillment of the covenant of the rainbow. (see *The Covenant*).

---

[186] Ether 13:2-4
[187] Ether 13:3

God preserved every kind of land animal and bird on Noah's Ark. Descendants of the original pairs ended up on different continents, thousands of miles apart. Even today, these diverse descendants can still breed with each other. For example, the descendants of the first cats can still breed...cougars with leopards, lions with tigers, and wild cats with domestic cats.

Each kind of plant and animal had the capacity to produce offspring with different designs to suit them for different environments. None of that variety was put there by humans. The information was there, programmed in their God-created DNA, right from the start.

Figure 41: Noah's Ark on Mount Ararat, Simon de Myle, 1520

AFTER THE FLOOD

As life refilled the earth after the Flood, continuing catastrophes buried snapshots of the fleeting environments that rose and fell. The fossil record shows some of the striking variety among the descendants of the animals that left Noah's Ark. For instance, we find fossils of more than 150 different species that arosnoe within two centuries after the first horses left Noah's Ark. Modern descendants of the first horses include zebras, donkeys, and stallions.

As Noah's descendants began to scatter across the world, some of them would have found themselves in difficult environments. As they still do in parts of the world today, caves would have made suitable homes for people who didn't have the ability or resources to build homes. These "cavemen" lived at the same time as the city builders and the hunter-gatherers.

We are still living with the consequences of the great flood. What some "experts" insist is global warming, or the new moniker "climate change" is actually the earth trying to find a new equilibrium. As the great ice sheets have melted away in the north, the land masses continue to rise, causing changes in the Great Lakes. The plants and animals continue to change, migrate, and adjust as the earth recovers from what was a traumatic experience.

> "In the days of Noah, God destroyed the world by a flood, and He has promised to destroy it by fire in the

last days: but before it should take place, Elijah should first come and turn the hearts of the fathers to the children..."[188]

Noah's influence extended all the way to the great patriarch Abraham. Abraham was just ten generations from Noah, being a direct descendant of Shem, (Noah's son), the father of all the "Semitic" peoples. Due to the duration of their lives, when Abraham was born, Shem was 390 years old, and his father Noah was 892 years old. Abraham was 58 years old when Noah died. These are important facts, since Abraham spent many years in the house of Noah and Shem, and received information directly from them (and was likely given the priesthood by Shem, as Melchizedek). He learned all the details about the flood from the very men who built the Ark and survived the flood (Noah and Shem). Noah knew Methuselah for many hundreds of years, who in turn knew Adam for many hundreds of years, which means that Abraham received reliable information about everything that happened since the very first day of creation.

Noah lived 350 years after the flood and died at the great age of 950. Many years later, God suggested that the righteous man he

---

[188] Teachings of the Prophet Joseph Smith, 337

used to save mankind was one of the three most righteous people to have lived up to that time[189].

One question that frequently is considered relates to the population of the earth. After the flood, how could eight people possibly have been responsible for the billions of people who are alive today. Rather than rely on an emotional analysis, in this case, we can simply do the math.

We start with Adam and Eve, one male and one female. We can assume that they marry and have children and that their children marry and have children. We will say that the population doubles every 150 years. After 150 years there will be four people, after another 150 years there will be eight people, after another 150 years there will be sixteen people, and so on. This growth rate is actually very conservative. In reality, even with disease, famines, and natural disasters, the US Census Bureau says that the world population currently doubles every 40 years or so.

After 32 of these periods, which is only 4,800 years, the world population would have reached almost 8.6 billion. That's more than the current population of 6.5 billion people by about 2 billion.

---

[189] Ezekiel 14:14, 20

But, we know from scripture, that around 2500 BC (4,500 years ago) the worldwide Flood reduced the world population to eight people. If we use our estimate that the population doubles every 150 years, starting with only Noah and his family in 2500 BC, 4,500 years is more than enough time for the present population to reach 6.5 billion.

If we believed the evolutionist version of history, and still used our conservative doubling period of 150 years, things really become unbelievable. Taking their history, and assuming humans have been around for 50,000 years, we would have 332 doubling periods. The world's population would now be an unbelievable number, a one followed by 100 zeros:

10,000,000,000,000,000,000,000,000,000,000,000,000, 000,000,000,000,000,000,000,000,000,000,000,000,00 0,000,000,000,000,000,000 people.

That's a lot of people, and a lot more than there are on the earth now, and that's with our ultraconservative doubling period of 150 years, not the 40 years we observe today. Even math supports the reality of the scriptures.

Our world now appears to be careening toward a crash. Most of us are troubled, if not frightened. Natural disasters; epic floods, tsunamis, hurricanes, a massive increase of seismic

AFTER THE FLOOD

activity and earthquakes, widespread famine and plagues abound upon the earth, and human wickedness that resembles the depraved days of Noah. Moral corruption and decay are all around us and nothing makes much sense any longer. However, we must stand firm in our faith and be ready.

The period of time immediately before the Messiah's return is sometimes called the time when the "footsteps of the Messiah" can be heard, ikvot meshicha (עִקְבוֹת מָשִׁיחַ). This is the time appointed by God for final redemption and the close of the present age. The condition of the world during the end of days will be grossly evil[190]. The world will undergo various forms of tribulation, called the "birth pangs of the Messiah"[191]. These birth pangs are said to last for 70 years, with the last 7 years as the most intense period of tribulation, the "Time of Jacob's Trouble" (עֵת־צָרָה הִיא לְיַעֲקֹב)[192].

The first wave of trouble comes from Edom (likely Europe); the second from Ishmael (the Arab states). Matthew's account of the Savior's sermon on Olivet foretold of these calamities that will precede His Second Coming[193]. We must be ready. We must remember. We must stand.

---

[190] 2 Peter 3:3; 2 Thessalonians 2:3-4, 2 Timothy 3:1-5
[191] Sanhedrin 98a; Ketubot, Bereshit Rabbah 42:4, Matthew 24:8
[192] Jeremiah 30:7
[193] Matthew 24-25

We are living in the last days. The words of Paul, hearken back to things we have learned about the world prior to the flood:

> "... in the last days perilous times shall come.
>
> "For men shall be lovers of their own selves, covetous ... blasphemers, disobedient to parents ... unholy,
>
> "Without natural affection ... incontinent, ...
>
> "... lovers of pleasures more than lovers of God."[194]

But Paul was not speaking of Noah's day, he was speaking of ours. He continues with a phrase that relates specifically to the corruption of the rainbow in our day speaking of those who, "creep into houses, and lead captive silly women laden with sins, led away with divers lusts."[195]

This philosophy has even found its way into the religions of the world, in the interest of 'tolerance' or 'inclusion' they give away the covenants of Enoch and Noah with statements such as "... precise rules of Christian conduct should not necessarily apply to problems of sexuality."[196]

President Spencer W. Kimball said:

> "Our world is now much the same as it was in the days of the Nephite prophet who said: '... if it were not for the

---

[194] 2 Timothy 3:1–4
[195] 2 Timothy 3:6
[196] London—British Council of Churches

prayers of the righteous ... ye would even now be visited with utter destruction. ...'[197] Of course, there are many many upright and faithful who live all the commandments and whose lives and prayers keep the world from destruction."[198]

---

[197] Alma 10:22
[198] Spencer W. Kimball, "Voices of the Past, of the Present, of the Future", April 1971, General Conference

THE TOKEN OF THE BOW

MEAT FOR YOU

The commandment God gave to Adam was specific regarding his source of food:

> "And God blessed them, and God said unto them, Be fruitful, and multiply, and replenish the earth, and subdue it: and have dominion over the fish of the sea, and over the fowl of the air, and over every living thing that moveth upon the earth. And God said, Behold, I have given you every *herb bearing seed*, which is upon the face of all the earth, and every tree, in the which is

> the *fruit of a tree* yielding seed; to you it *shall be for meat.*"[199]

While the commandment given to Noah was similar, it was also different:

> "And God blessed Noah and his sons, and said unto them, Be fruitful, and multiply, and replenish the earth. And the fear of you and the dread of you shall be upon every beast of the earth, and upon every fowl of the air, upon all that moveth upon the earth, and upon all the fishes of the sea; into your hand are they delivered. *Every moving thing that liveth shall be meat for you;* even as the green herb have I given you all things."[200]

Just as God had authorized mankind to eat "herb bearing seed" many centuries earlier, after the flood, God gave His permission for mankind to eat "every moving thing", including all animals that move on the Earth and swim in the sea. It appears that laws regarding the eating of clean and unclean animals were not given until the Law of Moses[201]. Although a distinction was made between clean and unclean animals prior to the Flood[202], this distinction seems to have applied only to the matter of

---

[199] Genesis 1:28-29
[200] Genesis 9:1-3
[201] Leviticus 11, Deuteronomy 14:3-21
[202] Genesis 7:2-3

which animals were suitable for sacrifice, not for consumption.[203]

But, just because God apparently did not authorize man to eat animal flesh before the Flood, does not mean that mankind obeyed. It seems likely that there were some people who went beyond what God allowed, and ate various kinds of animals anyway. It is not difficult to imagine those living just prior to the Flood, whose every thought was evil continually[204], leaning over a sacrificial sheep, smelling the sweet aroma, and taking a piece for themselves.[205]

So why did Abel raise flocks, if he and his descendants were supposed to be vegetarians? Although the scriptures do not say exactly why Abel was a "keeper of sheep"[206], most likely it was because by raising sheep, Abel could provide clothing for himself and others, as well as provide animals to be offered in sacrifice. Sacrifice to cover sin was instituted by God when He clothed Adam and Eve with skins[207]. It was taught to Cain and Abel[208].

---

[203] Genesis 8:20
[204] Genesis 6:5
[205] 1 Samuel 2:12-17
[206] Genesis 4:2
[207] Genesis 3:21
[208] Genesis 4:4

In the revelation commonly known as the 'Word of Wisdom'[209] modern day insight is provided:

> "Yea, flesh also of beasts and of the fowls of the air, I, the Lord, have ordained for the use of man with thanksgiving; nevertheless they are to be used sparingly; And it is pleasing unto me that they should not be used, only in times of winter, or of cold, or famine."[210]

This scripture has been used to promote one perspective over another, and while I will not provide an analysis of the full revelation, some discussion of this portion seems fitting, considering what we have just learned about the Lord's instructions to Noah upon leaving the ark.

Although it cannot be treated as simply a guide to better health, Church leaders and members have consistently extolled this purpose of the Word of Wisdom. They have pointed to the fact that it was given for our "temporal salvation"[211], that it explicitly tells us what is good and not good for our bodies[212], and that it includes promises appearing to relate to physical health[213].

---

[209] Doctrine & Covenants 89
[210] Doctrine &Covenants 89:12–13
[211] Doctrine & Covenants 89:2
[212] Doctrine & Covenants 89: 7–16
[213] Doctrine & Covenants 89:18, 20

However, there might also be a danger in assuming it is primarily a health code that will be unequivocally confirmed by scientific research.[214] If we believe the two are inextricably linked, the danger is that when scientific assertions seem to contradict the counsel in the Word of Wisdom, our loyalty to it might diminish, even if the science later proves to be wrong. If science tells us caffeine is bad for our health, this might strengthen our resolve to abstain from tea and coffee. But what happens when science uncovers beneficial aspects to caffeine or links the consumption of tea, coffee, and even alcohol to positive health benefits? Experts leading the small but growing interest in low-carb and so-called "Paleo" diets marshal their own lines of evidence to assert the health benefits of meat consumption, sometimes even at dramatically high levels. If we believe there is "scientific proof" that consuming more meat is good for us, our commitment to the Word of Wisdom as a health code could cause on us to rethink our interpretation and implementation of related verses.

One clue to the meaning of verse 13 might lie in the sole reason the Lord gives in the verse itself for abstaining from meat except during certain times. The Lord says it is "pleasing" to him.

---

[214] Paul Y. Hoskisson, The Word of Wisdom in Its First Decade," *Journal of Mormon History* 38/1 (Winter 2012): 140

Hugh Nibley suggests that the use of the word sparingly in D&C 89:12 means "sparing God's creatures." He goes on to say, "The family who needs a deer to get through the winter have a right to that. The Lord will not deny them, but he is also pleased with those who forbear."[215]

Apostle Lorenzo Snow said, "We have no right to slay animals or fowls except from necessity, for they have spirits which may some day rise up and accuse or condemn us."[216] Apostle Joseph Fielding Smith explained, "Although there was no sin in the shedding of their blood when required for food ... to take the life of these creatures wantonly is a sin before the Lord. It is easy to destroy life, but who can restore it when it is taken?"[217]

Clearly the Lord ordained the use of animal flesh for human consumption. He has warned against forbidding the use of "meats" and explained that they are part of the abundance of this earth with which he has blessed his children. While it is clear that Section 89 includes the admonition to use the flesh of animals "sparingly," it is interesting that the counsel to use it only during times of winter, cold or famine is prefaced simply with the explanation that this is "pleasing" to the Lord.

---

[215] Hugh Nibley, "Word of Wisdom: Commentary on D&C 89"
[216] Dennis B. Horne, ed., *An Apostle's Record: The Journals of Abraham H. Cannon* (Clearfield, UT: Gnolaum Books, 2004), 424
[217] Joseph Fielding Smith, "Is It a Sin to Kill Animals Wantonly?" *Improvement Era*, August 1961, 568

MEAT FOR YOU

One thing we can know for sure is that before the Flood, we never hear God granting permission to humans to eat animal flesh. Yet, at least three times prior to the Flood the Bible mentions God authorizing the fruit of the earth for man's consumption. Furthermore, Genesis 9:2-3 stresses that after the Flood a vastly different relationship existed between animals and humans. Now animals would fear humans, since humanity was permitted to use the flesh of animals for food, "even as the green herbs" were permitted since the beginning of the creation[218]. We also have dietary laws describing in minute detail the eating of animals. These include the accounts in Leviticus 10, 11 and 20, as well as Deuteronomy 14, Judges and Ezekiel. It becomes clear that not only is the eating of meat allowed, it has now become part of God's command to His children, "every living moving thing that liveth shall be meat for you."

To reinforce the point, Paul had this to say about the latter days:

> "... in the latter times some shall depart from the faith, giving heed to seducing spirits, and doctrines of devils; Speaking lies in hypocrisy; having their conscience

---

[218] Genesis 1:29, 9:3

seared with a hot iron; Forbidding to marry, and commanding to abstain from meats. ..."[219]

In another attempt to turn God's commandments on their head, Satan will lead some to depart from the faith in the latter-days. Isn't it interesting that among the lies he will use are, forbidding to marry, and refusing to eat meat.

---

[219] 1 Timothy 4:1-3

BAPTISM OF THE EARTH

Concerning the baptism of the earth, President Brigham Young said, as only he could:

> "It has already had a baptism. You who have read the Bible must know that that is Bible doctrine. What does it matter if it is not in the same words that I use, it is none the less true that it was baptized for the remission of sins. The Lord said, `I will deluge (or immerse) the earth in water for the remission of the sins of the people'; or if you will allow me to express myself in a familiar style, to kill all the vermin that were nitting, and breeding, and polluting its body; it was cleansed of its filthiness; and soaked in the water, as long as some of our people ought to soak. The Lord baptized the earth for the remission of sins, and it has been once cleansed for the filthiness that

has gone out of it, which was in the inhabitants who dwelt upon its face."[220]

As early as 1835 there were religious leaders teaching that Noah's flood was, in fact, a baptism of the earth. William Trollope wrote, "The preservation of Noah and his family in the ark from perishing by water is emblematic of baptism, inasmuch as it is only by baptism that Christians are admitted into the Church."[221]

While the earth may not be considered to have "sinned", it nonetheless was a place where the sinful acts of its inhabitants occurred. So, while we may not understand the full necessity, in my mind we have only to look at the Savior to realize that simply viewing baptism as a necessary ordinance for the cleansing of sin, ignores many other, equally significant implications of this saving ordinance (see my book *Most Glorious of All* for a broader discussion on this topic). It is sufficient to know that those who have considered the question, including prophets and revelators have come to the conclusion that not only was the baptism of the earth a very real aspect to the flood, but a necessary one, which will ultimately be followed by a baptism of fire.

---

[220] Journal of Discourses 1:274
[221] William Trollope, *Analecta Theologica: A Digested and Arranged Compendium of the Most Approved Commentaries on the New Testament* , 1835, 2:625

"Thus, modern revelation teaches that God indeed suffered great sorrow over the Flood, which served as the baptism of the earth."[222]

"Some people talk very philosophically about tidal waves coming along. But the question is—How could you get a tidal wave out of the Pacific ocean, say, to cover the Sierra Nevadas? But the Bible does not tell us it was a tidal wave. It simply tells us that "all the high hills that were under the whole heaven were covered. Fifteen cubits upwards did the waters prevail; and the mountains were covered." That is, the earth was immersed. It was a period of baptism."[223]

"Another great change happened nearly two thousand years after the earth was made. It was baptized by water. A great flow of water come, the great deep was broken up, the windows of heaven were opened from on high, and the waters prevailed upon the face of the earth, sweeping away all wickedness and transgression-a similitude of baptism for the remission of sins. God requires the children of men to be baptized. What for? For the remission of sins. So he required our globe to be baptized by

---

[222] Joseph B. Romney, "Noah, The Great Preacher of Righteousness," *Ensign*, Feb 1998, 22
[223] John Taylor, *Journal of Discourses* 26:74-75

a flow of waters, and all of its sins were washed away, not one sin remaining."[224]

"This earth, in its present condition and situation, is not a fit habitation for the sanctified; but it abides the law of its creation, has been baptized with water, will be baptized by fire and the Holy Ghost, and by-and-by will be prepared for the faithful to dwell upon."[225]

---

[224] Orson Pratt, August 1, 1880 *Journal of Discourses* 21:323
[225] Brigham Young, June 12, 1860 *Journal of Discourses* 8:83

BAPTISM OF THE EARTH 247

248 THE TOKEN OF THE BOW

The Apotheosis of Washington, Constantino Brumidi, US Capitol

WHO AM I?

To understand the atonement, we must first understand the fall. And, to understand the fall we must understand creation. These events are inextricably tied to one another. Bruce R. McConkie, in a February 1981 devotional said:

> "The three pillars of eternity, the three events, preeminent and transcendent above all others, are the creation, the fall, and the atonement. These three are the foundations upon which all things rest. Without any one of them all things would lose their purpose and meaning, and the plans and designs of Deity would come to naught.
>
> "If there had been no creation, we would not be, neither the earth, nor any form of life upon its face. All things,

all the primal elements, would be without form and void. God would have no spirit children; there would be no mortal probation; and none of us would be on the way to immortality and eternal life.

"If there had been no fall of man, there would not be a mortal probation. Mortal man would not be, nor would there be animals or fowls or fishes or life of any sort upon the earth. And, we repeat, none of us would be on the way to immortality and eternal life."

He concluded his opening remarks, "And so I now say: Come and let us reason together; let us reason as did righteous men of old that we may come to understanding."

If there had been no atonement of Christ, all things would be lost. The purposes of creation would vanish away. Lucifer would triumph over men and become the captain of their souls. And, we say it again, none of us would be on the way to immortality and eternal life.

As part of understanding creation, we must understand our own inception. The age-old question, "Where do I come from?"

How is this relevant in a discussion about the token of the rainbow, or of the global flood? Only by understanding the inerrant nature of the account given us in scripture, the fact that these accounts are both literally, and figuratively true, can we hope to find a satisfying answer to these questions. The token is only important because of God's love for us. Our relationship with him is at the heart of our existence; past, present and future.

So, let's begin at the beginning. Creation, as we have seen in the previous chapter on the nature of light, was a proactive event. The creator of the world, Jehovah, spoke this creation into existence. It was a faithful, and a faith-filled act. Joseph Smith said:

> "When a man works by faith he works by mental exertion instead of physical force. It is by words, instead of exerting his physical powers, with which every being works when he works by faith. God said, 'Let there be light: and there was light.' ... And the Saviour says: 'If you have faith as a grain of mustard seed, say to this mountain, "Remove," and it will remove; or say to that sycamore tree, "Be ye plucked up, and planted in the midst of the sea," and it shall obey you.' Faith, then, works by words; and with these its mightiest works have been, and will be, performed. ...
>
> "... The whole visible creation, as it now exists, is the effect of faith. It was faith by which it was framed, and it is by the power of faith that it continues in its organized form, and by

which the planets move round their orbits and sparkle forth their glory"[226]

Prior to the set of events related in the Book of Moses, all existed only in the spiritual. We do not have a detailed account of our spiritual creation, just references to the fact that it occurred. Joseph Fielding Smith said, "The account of the creation of the earth as given in Genesis, and the Book of Moses, and as given in the temple, is the Creation of the physical earth, and of physical animals and plants. ... There is no account of the Creation of man or other forms of life when they were created as spirits."[227]

God does say, "For I, the Lord God, created all things, of which I have spoken, spiritually, before they were naturally upon the face of the earth... And I, the Lord God, had created all the children of men... for in heaven created I them."[228]

So, we see that part of this incredible spiritual creation, was you...and I...and everyone who has ever, or will ever, live. While bodies continue to be formed from the physical elements encompassed in the Lord's creation of this world, each of us existed spiritually long before this world ever came into physical being. Modern scripture reveals, "Man was also in the

---

[226] Lectures on Faith, 72–73
[227] Doctrines of Salvation, 1:75
[228] Moses 3:5

beginning with the Father"[229]. Through this revelation, we learn that we were created of something more eternal in nature. We were created as the offspring of Elohim, the Eternal Father. Deuteronomy 14:1 states that "Ye are the children of the Lord your God..."

And, just as physical offspring carry the characteristics of their physical parents, this pattern existed with our spiritual begetting. We carry the spiritual characteristics that make us God's children, having been created "in his own image."[230]

Tzelem Elohim, the "image of God", can also be understood as "God's shadow" since the root tzel (צל), is the word for "shadow". In a sense, a copy, even if at times an imperfect one,

---

[229] Doctrine & Covenants 93: 23,29
[230] Genesis 1:27, Doctrine & Covenants 20: 18

of the perfect real image. Tzelem Elohim is the conjugation of two terms. Tzelem (צֶלֶם), is a Hebrew term which describes "to carve" or "to cut." In Daniel, chapter 3, it is used to refer to the statue that King Nebuchadnezzar, in an act of pride and idolatry, erected this image of himself for everyone to bow down to in the plain of Dura. Elohim (אֱלֹהִים) occurs 2,602 times in the Bible. It is the most common Hebrew word in the Bible translated as God. In 1919 the First Presidency stated that "God the Eternal Father, whom we designate by the exalted name-title 'Elohim,' is the literal Parent of our Lord and Savior Jesus Christ, and of the spirits of the human race." Bruce R. McConkie said this of Elohim:

"The Father's name is Elohim;

"Jehovah is his Son.

"Above all Gods they stand supreme,

"And rule the universe." [231]

In Genesis 1:27 this phrase is coupled with "created", or Demuth (דְּמוּת). Demuth, from the root meaning "to be like" is used to describe forbidden images of anything in heaven or on earth (Exodus 20:4). There is a sense of the object being

---

[231] Bruce R. McConkie, April 1977, General Conference

invested with an essence, carrying with it abstract attributes of the real, including self-consciousness, mercy, power, authority, intelligence or any of the attributes of God's nature. In ancient times it was thought that the gods, or rather their essence, inhabited the images that were carved as idols. In a sense, the one who owned the image exercised some authority over the god; he could fashion him out of wood or rock, move him about, or even diminish his authority by having many gods at the same time. God's command against making such objects sets at odds the idea that He can be manipulated by His creation, yet there is something of God that resides in every man.

So, in Genesis 1, the idea is clearly that man represents God, including His presence in the world and His authority. None of the other creations of God are in God's image.

You have undoubtedly heard this inate ability to be like our Hevenly Father referred to as "spiritual DNA." And while this is certainly a convenient metaphor, it hardly does it justice. For, DNA is subject to all the limitation of physical existence. There can be transcription errors, genetic disease, faulty sequences introduced by environmental factors, and so on.

There is a greek term, ἀποθέωσις, or apotheosis. This term literaly means "to deify" It is the glorification of a subject to a divine level. It makes sense that this has been a concept taught by everyone from the ancient Egyptians, the Mestopotamians, the Greeks and Romans. This belief even persisted in Ancient China. The Ming dynasty included an epic, *Investiture of the Gods* (Fenshen Yanyi), a novel from the 16th century that contained numerous deification stories. Each of these cultures remembered the ancient truth that we are here to become like our Heavenly Father, and in the absence of scripture, modified these stories, or incorporated them in their own narratives.

For many in these cultures, and those affected by them, the concept of deification has become watered down, considered heresy, and mingled with the "philosophies of men." These teachings try to trivialize men. This is part of his effort to deceive mankind into worshiping him, "because he had fallen from heaven, and had become miserable forever, he sought also the misery of all mankind."[232] This misery can only come about if we fail to acknowledge who we really are. God told Moses that he was his son. The greatest weapon Satan had was to remove that belief from him, to say to him "Moses, Son of man."[233]

> "Man is the child of God, formed in the divine image and endowed with divine attributes, and even as the infant son of an earthly father and mother is capable in due time of becoming a man, so the undeveloped offspring of celestial parentage is capable, by experience through ages and aeons, of evolving into a God"[234].

In the Psalm of Asaph, God says, "Ye are gods; and all of you are children of the most High."[235]

If God is to be believed, and we are His children, with His characteristics imprinted upon us, what does this mean?

Those who believe, and teach evolution, rely on several assumptions. One, that the Earth (and all creation) are billions of years old. And two, that death is the driving force to our existence. "The secrets of evolution are time and death: time for

---

[232] 2 Nephi 2:18
[233] Moses 1:12`
[234] *Improvement Era*, Nov. 1909, 81
[235] Psalms 82:6

the slow accumulation of favorable mutations, and death to make room for new species."[236] Millions, or billions of years are a necessity, and the death of our ancestors was simply a requirement for our improvement, or evolution. Evolution is a theory based on death. Contrast that to the gospel, and creation, which are based on life.

Historians who believe in evolution (and the millions and billions of years it requires) have an interesting perspective on man's progress. While man has been on the earth for millions of years, they hold that 2.6 million years ago, stone tools were first created followed by fire (2.3 million years ago), boats (900 thousand years ago), and on and on.

It is interesting that the next period, referred to as "Neolithic", began about 10,000 years ago. It is only here, in the last 10,000 years, by their estimation, that things really began to happen. In this time cities were built, copper was smelted (5,000 BC), the wheel was first used (4,000 BC), writing began (3,000 BC) and cast iron was developed (500 BC). While there are challenges with the specific dating of many of these discoveries, placing them after the great flood seems appropriate.

Since the time of Christ (and here the accuracy of dating begins to be more truly historic and reliable), woodblock printing (300

---

[236] Carl Sagan, *Cosmos*, "One Voice in the Cosmic Fugue."

AD), matches (577 AD), toilet paper (589 AD), algebra (900 AD), eyeglasses (1286 AD), the printing press (1439 AD), the telescope (1608 AD), the steam engine (1765 AD), and the cotton gin (1793 AD) to name a few.

In the 1800's things really began to take off, especially after the restoration of the gospel. In fact, British historian Paul Johnson wrote an amazing history book in 1991 which drew attention to this amazing period. This book, called "The Birth of the Modern," revealed this critical time that changed the course of world history. After 1830, the year the Church was organized in this dispensation, a partial list of achievements reads like a definition of the modern world: photography, reaper and cotton gin, telegraph, gas refrigeration, rubber tire, sewing machine, elevator, hypodermic needle, internal-combustion engine, dynamite, typewriter, automobile, phonograph, light bulb, airplane, radio and television, anesthesia and Novocain, air conditioning, nuclear reactor, xerox machine, fax machine, computers, and the Internet.

While to many this seems miraculous, in light of our understanding, it should be no surprise. As we have already discussed in this volume, there is little credibility to the dating systems used by archeologists, they are too contradictory. Since the flood, and the repopulation of the world, man's progress has been largely continuous and impressive. While there have been setbacks, from the tower of babel, the plagues of Europe, and several world wars, the progress has been nonetheless consistent in its onward push.

It can be difficult to recognize the achievements of earlier generations due to the biases and agendas of those creating our written histories. We typically think of ancient peoples as ignorant, savage and uncivilized. None more so than the Native Americans, descendants of Lehi and other ancestors, who have been vilified and trivialized to justify the elimination of their culture and the taking of their land. However, these children of God were the inventors of the syringe, baby formula, cigars, petroleum mining, popcorn, potatoes and chocolate. Not to mention their cultural contributions, including the Iroquois Confederacy, which influenced Benjamin Franklin, and thereby the creators of the American form of government, the "more perfect union". In addition, over 2,200 words from native languages have made their way into the English language. Our myopic view of history robs us of a valuable perspective on who we are, and what we have accomplished with the birthright we have been given.

If we are who Heavenly Father says we are, why would we be surprised? With this seed of godhood in us, surely we have capacities we do not understand, and cannot comprehend. In fact, our failure to achieve great things would be inconsistent with this heritage. What we have achieved, in spite of our lack of understanding, and considering the distractions, temptations and confusion Satan throws at us says much more about our capacity, than it does about our ability. Biologically, we are not much more complicated than most of God's other creations. What makes us unique is that we were created "in the image of God." This likeness instills in us an innate ability to be like Him. And who is He? The Creator of the Universe. The Master Builder. The Master Teacher. All of these characteristics are ours, if we embrace them.

Scriptures say that the natural man is an enemy to God. A natural man is one who chooses to be influenced by passions, desires, appetites, and senses of the flesh rather than by the promptings of the Holy Spirit. The natural man can comprehend physical things but not spiritual things. In spite of our weaknesses, through the Atonement of Jesus Christ we can still love, serve, learn and build. Our history shows the greatness of what Heavenly Father has instilled in us. As we put off the natural man, our capacity will only grow. We will achieve anything.

As you consider the great inventions that continually pour out on the world, see them for what they are, a testimony of the greatness within each of us. A sign of our divine heritage.

The Savior Himself said, "Is it not written in your law, I said, Ye are gods?"[237]

---

[237] John 10:34

CONCLUSION

This book had several goals. I hoped to help readers understand that the Bible is a true history. I hoped to remind readers of the great love God has for his children, collectively and individually. I hoped to provide a solid footing for anyone wishing to embrace the truth of God's covenants with His people. I hoped to help people realize that the flood came as people rejected their families. And, I hoped to reclaim the token of the rainbow from Satan's attempts to confuse and distract from its true meaning.

For some, a scriptural analysis of the rainbow and its meaning may have been sufficient. But, as the world has become more and more complex, the methods used by Satan to distract, discredit and corrupt have become more insidious. We must arm ourselves with truth. It is not enough to simply say that the rainbow does not mean what much of the world wishes it to mean. And, if the rainbow is a token of a covenant between God and each of us, we must come to terms with its foundational event, the global flood. Without a flood, without a world-wide destruction of all flesh, without the love and concern shown by a loving Heavenly Father to save eight righteous individuals from this global flood, what is the basis for the covenant the Lord made first with Enoch, and then with Noah, and finally with each of us?

We cannot pick and choose where we are going to be faithful, and where we are going to trust the Lord. It is no wonder so many young people today struggle with faith. Without absolute trust in the Lord, there is no faith, and no strength.

The rainbow has not simply been misused. Like each of Satan's attempts to defile sacred symbols and messages, he seeks to stand them on their head. The corruption of the rainbow has been purposeful. As you have seen in this book, the flood came as people hated their own blood. As the granddaughters of Noah sold themselves to the world, Noah repented, and God sent destruction…to save the children who would follow. Satan is attempting to use the rainbow to bring about the same destruction of the family that prefaced its first appearance.

President Marion G. Romney said: "The bringing of peace requires the elimination of Satan's influence. Where he is, peace can never be. Further, peaceful coexistence with him is impossible… He promotes nothing but the works of the flesh…As a prelude to peace, then, the influence of Satan must be completely subjugated…"[238]

The works of the flesh ignore the truth of God's word. We, as Latter-day Saints know what happens when a people do not have God's word. In the absence of scripture, which can come either through its physical loss, or through a failure to believe its truths, apostasy follows. We must arm ourselves, and our youth with truth, not excuses. The scriptures are either true or they are not. The global flood either happened or it did not. If it did not, then all the references to it, by Joseph Smith, the Apostle Peter, and others, including the Savior Himself, are untrue. If it is true, then all the promises made by the Lord are true as well.

We cannot pick and choose. And we cannot continue to allow Satan to corrupt sacred tokens. We must learn their truths, and then stand. This book is my attempt to contribute to the telling of these truths, and the reclaiming of the wonderful token of the bow.

President Brigham Young said, "Brethren and sisters, I wish you to continue in your ways of welldoing; I desire that your minds may be opened more and more to see and understand things as they are. This earth, in its present condition and

---

[238] The Price of Peace, *Ensign*, Oct. 1983, 4, 6

situation, is not a fit habitation for the sanctified; but it abides the law of its creation, has been baptized with water, will be baptized by fire and the Holy Ghost, and by-and-by will be prepared for the faithful to dwell upon."[239]

Author's Note: Thank you to all those who helped make this book possible. The support provided by my wife, Carol and my family and friends were invaluable. A special thank you to Mark and Esther Stringer, Jeff and Helen Christensen, and Irene and Tomasi Tukuafu.

Since this is a book that, at its core, is about family, it is fitting that the art used is the creation of family. Both family by blood, and family by preference. You have all given of yourselves and your talent, and I thank you.

R. Lane Wright

---

[239] *Journal of Discourses*, 8:83

INDEX

Abel, 9, 237
Abraham, 9, 145, 146, 147, 162, 208, 228
Abram, 9, 208
Adam, 1, 3, 4, 7, 8, 26, 27, 30, 41, 53, 65, 143, 144, 145, 146, 147, 152, 228, 229, 235, 237
Adam ondi Ahman, 27
Alexander the Great, 104
Andes, 160
Angkor, 116
apocrypha, 4
apotheosis, 256
Archaeopteryx lithographica, 111
Archangel Michael, 102
Aristotle, 104, 105, 191
ark, ii, 18, 21, 23, 25, 26, 28, 29, 30, 31, 34, 35, 36, 37, 38, 39, 40, 41, 43, 44, 48, 51, 52, 67, 69, 128, 134, 157, 163, 164, 173, 244
Australia, 116, 157, 158
Aztec, 159
Babylon, 155
baptism, 140, 145, 243, 244, 245
barge, 32
Basal Sauk Megasequence, 82
behemoth, 107
bird, 36, 68, 70, 99, 111, 160, 226
Bishop Bell, 115
boat, 23, 25, 32, 35, 155, 156, 159, 161
Book of Mormon, 172, 173
Book of Moses, 8, 11, 212, 215, 252
bridegroom, 55

Brigham Young, 209, 243, 246
Brother of Jared, 38, 48
Bruce R. McConkie, 249, 254
cagar, 52
Cain, 237
carbon-14, 95
Carl Sagan, 258
Chaldean, 156
challown, 35
Charles Darwin, 86
Charles Kingsley, 104
China, 116, 155, 256
Chinese, 102, 116, 151, 155
Christ, 19, 30, 55, 69, 128, 129, 138, 145, 162, 163, 164, 168, 169, 175, 185, 186, 188, 213, 216, 250, 254, 258, 261
City of Enoch, 131, 214, 215, 216
cloudbows, 197
corruption, 29, 128, 231, 232, 264
covenant, ii, iii, 8, 10, 11, 12, 13, 17, 21, 24, 44, 69, 120, 140, 143, 144, 145, 146, 147, 148, 149, 168, 207, 208, 209, 210, 211, 212, 213, 214, 216, 263
cratons, 62, 65
cubit, 25, 30, 36, 37, 39, 46, 52
curse, 4, 17
David, 43, 44, 168, 169, 173
Delaware Indians, 160
deluge, i, 56, 164, 222, 243
Demuth, 254
Dieter F. Uchtdorf, 141
dinosaur, 76, 92, 103, 106, 107, 109, 114
Dinosauria, 109

DNA, 40, 41, 84, 85, 86, 92, 93, 226, 255
door, 25, 38, 41, 48, 49, 51, 52, 54, 55, 56, 57, 67, 136
dove, 36, 63, 67, 68, 69, 70, 154, 156, 159
Dr. Storrs L. Olson, 96
dragon, 102, 103, 106, 107, 108, 114, 116
drought, 124, 125
E=mc², 186
Eden, 27, 30, 37
Egypt, 25, 29, 30, 54, 92, 119, 120, 123, 126, 207, 208
Egyptian, 30
Einstein, 186
Elohim, 8, 253, 254
Enoch, 1, 3, i, 2, 3, 4, 5, 6, 8, 9, 10, 17, 19, 131, 132, 133, 134, 137, 138, 152, 162, 211, 212, 213, 214, 215, 216, 225, 232, 264
Eve, 41, 143, 144
evolution, 40, 83, 88, 91, 92, 257, 258
Ezekiel, 123, 141, 192, 197, 229
Fabrizio Carbone, 182
families, iii, 10, 21, 138, 147, 149, 159, 161, 175, 205
family. *See* Families
famine, 4, 163, 192, 231
feathers, 96
fogbows, 197
fossil, 26, 28, 32, 62, 65, 73, 74, 78, 81, 82, 90, 102, 227
fountains of the deep, 60, 62, 63
Gabriel, 7
garnet, 37, 38
Geographia, 102
George Harlow, 38
geshem nedavot, 122
giants, 15, 16
gibborim, 15, 16
gigantes, 16

Gilbert Baker, 202
global flood, 78, 87, 92, 153, 164, 250, 264, 265
global warming, 227
gopher wood, 25, 26, 29
Grand Canyon, 62, 80, 81, 82, 223
Greece, 29, 158
Grimm Fairy Tale, i
Ham, 7, 10, 41, 134
hand cart trek, 134
Harold Arlen, 190
Heavenly Father, 128, 129, 148, 149, 176, 205, 210, 256, 261, 264
Holy Spirit, 68, 70, 261
Honi, 124, 125, 126, 127
Hunt-Lenox Globe, 101
Hyman Arluck, 190
Illinois, 2, 28
Inca, 160
inclusion, 232
India, 62, 104, 157
International Order of Rainbow for Girls, 202
Iroquois Confederacy, 260
Isaac Newton, 194
Isaiah, 9, 55, 107, 173
Isidore Hochberg, 190
Israel, 15, 24, 119, 123, 125, 168, 169, 173, 190
Jacob, 2, 146
Japeth, 7, 10, 41, 134
Jaredite's, 52
jelly-fish, 78
Jericho, 55
Jerusalem, 9, 55, 126, 215, 216
Job, 107, 222
John Taylor, 5, 127, 128, 181, 245
Joseph, 5, 6, 7, 11, 20, 25, 34, 121, 122, 131, 134, 147, 162, 163, 179, 180, 181, 182, 192,

211, 212, 228, 245, 251, 252, 265
Joseph Fielding Smith, 7, 12, 25, 252
judgement, 5, 20, 21
Kermit, 189, 199
King Arthur, 105
King of Salem, 9
kopher, 29, 32, 33
Kronosaurus, 107, 108
Lamech, 5, 6
leviathan, 107
Leviathan, 107
Light of Christ, 175
Linda Curley Christensen, 198
Lord Kelvin, 86, 87
Lot, 54
macroevolution, 83
Marco Polo, 116
Marion G. Romney, 264
Mark Twain, 87
meat, 236, 241, 242
Melchizedek, 9, 10
Messiah, 163, 192, 231
Methuselah, 2, 3, 4, 5, 18, 19, 53, 54, 60, 133, 134, 228
Mexico, 159
microevolution, 84
Midrash, 53, 192
Mississippi River, ii
Missouri, 27, 28
Moses, 3, 7, 8, 11, 15, 16, 17, 19, 20, 21, 24, 25, 30, 32, 52, 89, 99, 131, 132, 133, 144, 145, 147, 148, 153, 162, 173, 179, 197, 213, 252, 257
NASA, 84, 85
National Geographic, 96, 97, 98
National Museum of Natural History, 96
Native Americans, 260
Natural disasters, 230
Neal A. Maxwell, 128
Neolithic, 258

Nephi, 25, 49, 52, 127, 162, 172, 257
nephilim, 15, 16
Nile, 119
Noah, 1, 3, i, ii, i, 1, 2, 3, 4, 5, 6, 7, 8, 9, 10, 11, 12, 13, 15, 16, 17, 18, 19, 20, 21, 23, 25, 26, 29, 30, 31, 32, 33, 35, 36, 37, 38, 41, 43, 51, 52, 53, 56, 57, 59, 64, 65, 67, 69, 108, 120, 122, 123, 127, 128, 129, 133, 134, 137, 138, 143, 146, 147, 152, 157, 162, 163, 164, 168, 173, 192, 205, 207, 210, 211, 212, 213, 214, 216, 222, 226, 227, 228, 244, 245, 264
oats, 6
Ojibwe, 159, 160
olive tree, 67, 70
Origin of Species, 86
Pangea, 26, 64, 65, 66
paraconformity, 78
Paul, 162, 164, 212, 232, 241, 259
Paul Harvey, 212
Paul Johnson, 259
Permineralization, 75
Peru, 114
Petrification, 74
Petrified, 74
Phison, 37
piercing, 35, 103
pitch, 25, 29, 32, 33, 156
polystratic, 77
population, 229, 230
Priesthood, 5, 7, 209
prism, 194
Psalm of Asaph, 257
Ptolemy, 102
rainbow, i, 161, 168, 184, 191, 192, 193, 194, 195, 196, 198, 199, 201, 202, 204, 205, 210, 211, 214, 250, 263
raven, 36, 67, 68, 69, 70, 156

Red Dragon, 105
Reflection, 193
Refraction, 193
René Descartes, 194
Robert Marshall, 198
Rodinia, 26, 61, 62, 63, 64, 65, 66
sacrifice, 140, 162, 237
Saint George, 102
Satan, 2, 68, 205, 242, 257, 261, 263, 264, 265
sauropod, 92, 107, 114, 115
Scott Kelly, 84
semantron, 29
Septuagint, 16, 26, 29
serpent, 106, 144
Shem, 7, 9, 10, 41, 134, 152, 228
Sir Richard Owen, 103, 109
Smithsonian, 96
Sodom, 54
Solomon's Temple, 43
Somewhere Over the Rainbow, 190
Son of Man, 132, 141, 163, 213
sons of God, 8, 10, 11, 12
sons of men, 8, 10, 11, 12, 18, 19, 20
Spencer W. Kimball, 23, 164, 232, 233
subduction, 63
subway garnet, 38
Tabernacle, 45, 173

tal, 122, 197
Talmud, 36, 54, 121, 122, 124, 193
tannin, 106, 107
Tanzania, 154
Tebah, 32
Tel Dan stela, 169
The Book of Jasher, 9
The Rainbow Connection, 190
Theodoric of Frieberg, 193
Thomas S. Monson, 142
Title of Liberty, 208
token, ii, iv, 163, 179, 192, 196, 205, 207, 208, 210, 211, 213, 214, 216, 250, 263, 265
tolerance, 232
Toltec, 159
trilobite, 89, 90
tsohar, 34, 36
Tzelem, 253
United States, 2, 159
universal flood, 11, 12
Utah, 114
vegetarian, 237
Wavelength, 195
wheat, 6
window, 25, 34, 35, 36
Wizard of Oz, 190
Yip Harburg, 190
Yom, 184
Zion, 1, 131, 133, 134, 214, 216